Securing AI Model Weights

Preventing Theft and Misuse of Frontier Models

SELLA NEVO, DAN LAHAV, AJAY KARPUR, YOGEV BAR-ON,
HENRY ALEXANDER BRADLEY, JEFF ALSTOTT

This document was revised in June 2024 to add acknowledgments, correct formatting, and make an addition to Appendix A.

For more information on this publication, visit **www.rand.org/t/RRA2849-1**.

About RAND

RAND is a research organization that develops solutions to public policy challenges to help make communities throughout the world safer and more secure, healthier and more prosperous. RAND is nonprofit, nonpartisan, and committed to the public interest. To learn more about RAND, visit www.rand.org.

Research Integrity

Our mission to help improve policy and decisionmaking through research and analysis is enabled through our core values of quality and objectivity and our unwavering commitment to the highest level of integrity and ethical behavior. To help ensure our research and analysis are rigorous, objective, and nonpartisan, we subject our research publications to a robust and exacting quality-assurance process; avoid both the appearance and reality of financial and other conflicts of interest through staff training, project screening, and a policy of mandatory disclosure; and pursue transparency in our research engagements through our commitment to the open publication of our research findings and recommendations, disclosure of the source of funding of published research, and policies to ensure intellectual independence. For more information, visit www.rand.org/about/research-integrity.

Published by the RAND Corporation, Santa Monica, Calif.

Library of Congress Cataloging-in-Publication Data is available for this publication.

ISBN: 978-1-9774-1337-6

Cover: WenPhoto/Adobe Stock, design by Haley Okuley/RAND.

About This Report[1]

As frontier artificial intelligence (AI) models—that is, models that match or exceed the capabilities of the most advanced AI models at the time of their development—become more capable, protecting them from malicious actors will become more important. In this report, we explore what it would take to protect the learnable parameters that encode the core capabilities of an AI model—also known as its *weights*—from a range of potential malicious actors. If AI systems rapidly become more capable over the next few years, achieving sufficient security will require investments—starting today—well beyond what the default trajectory appears to be.

We focus on the critical leverage point of a model's weights, which are derived by training the model on massive datasets. These parameters stem from large investments in data, algorithms, compute (i.e., the processing power and resources used to process data and run calculations), and other resources; compromising the weights would give an attacker direct access to the crown jewels of an AI organization's work and the nearly unrestrained ability to abuse them.

This report can help information security teams in frontier AI organizations to update their threat models and inform their security plans, as well as aid policymakers engaging with AI organizations in better understanding how to engage on security-related topics.

Meselson Center

RAND Global and Emerging Risks is a division of RAND that delivers rigorous and objective public policy research on the most consequential challenges to civilization and global security. This work was undertaken by the division's Meselson Center, which is dedicated to reducing risks from biological threats and emerging technologies. The center combines policy research with technical research to provide policymakers with the information needed to prevent, prepare for, and mitigate large-scale catastrophes. For more information, contact meselson@rand.org.

Funding

Funding for this work was provided by gifts from RAND supporters.

Acknowledgments

We thank the following experts, as well as those who did not opt to be named, for their insights and contributions to the ideas in this report: Vijay Bolina, Paul Christiano, Jason Clinton, Lisa Einstein, Mark Greaves, Chris Inglis, Geoffrey Irving, Eric Lang, Jade Leung, Tara Michels-Clark, Nikhil Mulani, Anne Neuberger, Ned Nguyen, Nicole Nichols, Chris Rohlf, Wim van der Schoot, Phil Venables, and Heather Williams. Not all consulted experts support all recommendations and conclusions in the report, and the views expressed in

[1] Author affiliations are as follows: Sella Nevo, Ajay Karpur, Jeff Alstott, Yogev Bar-On, and Henry Bradley (RAND), and Dan Lahav (Pattern Labs). This is the final research report following an interim report that was published in 2023: Sella Nevo, Dan Lahav, Ajay Karpur, Jeff Alstott, and Jason Matheny, "Securing Artificial Intelligence Model Weights: Interim Report," RAND Corporation, WR-A2849-1, 2023.

this report do not necessarily represent the views of any institution or organization with which the consulted participants are affiliated, including the U.S. government.

We thank the following people for their contributions to the editing, reviewing, and strategic thinking for this report: Arwen Bicknell, Lennart Heim, Chad Heitzenrater, Holden Karnofsky, Gabriel Kulp, Jason Matheny, James Torr, Mary Vaiana, N. Peter Whitehead, Henry Willis, and Li Ang Zhang. And finally, we are grateful to our peer reviewers, Josh Wallin at the Center for New American Security and Michael Vermeer at RAND, for their insightful and thoughtful comments and critique of the report.

Summary

The goal of this report is to improve the security of frontier artificial intelligence (AI) or machine learning (ML) models. (*Frontier* models are those that match or exceed the capabilities of the most advanced AI models at the time of their development.) Our analysis focuses on foundation models, and specifically large language models and similar multimodal models. We focus on the critical leverage point that is the core of a model's intelligence and capabilities: its *weights*, a term used here to refer to all learnable parameters derived by training the model on massive datasets. These parameters stem from large investments in data, algorithms, compute (i.e., the processing power and resources used to process data and run calculations), and other resources; compromising the weights would give an attacker direct access to the crown jewels of an AI developer's work and the ability to exploit them for their own use.

While many existing information security frameworks lay strong foundations for developing security plans, the growing policy discussions and public interest in preventing frontier model misuse and theft have highlighted the need for a shared language between AI developers and policymakers to foster a mutual understanding of threat models, security postures, and security outcomes, grounded in the same technical definitions. To advance such understanding, we offer four key contributions, detailed in Figure S.1.

FIGURE S.1
Key Contributions of This Report

We identify approximately 38 meaningfully distinct attack vectors.

In most cases, an organization's vulnerability to just one of these vectors can compromise its security. We provide hundreds of real-world examples in which these attack vectors were deployed successfully, demonstrating that they are feasible and providing context on what such attacks look like in practice.

We explore a variety of potential attacker capabilities, from opportunistic (often financially driven) criminals to highly resourced nation-state operations.

This categorization of attacker capabilities allows organizations to identify sequential priorities depending on their current security infrastructure.

We estimate the feasibility of each attack vector being executed by different categories of attackers.

About a dozen attack vectors are likely infeasible for nonstate actors, but they are feasible for state actors, highlighting the need for significantly more-capable security systems to defend against state actors. Expert opinions vary significantly on the capabilities of state actors and how to defend against them.

We propose and define five security levels and recommend preliminary benchmark security systems that roughly achieve the security levels.

Each level is defined as being secure against attack vectors feasible for increasingly capable categories of malicious actors. The benchmarks can help to calibrate the trade-off between security investment and protection against different actors. The security levels are not meant to be used as a standard. Rather, they provide concrete suggestions for steps that frontier AI organizations can take at different stages of their continuous security enhancement strategy.

Recommendations

Avoiding significant security gaps is highly challenging and requires comprehensive implementation of a broad set of security practices. However, we highlight several recommendations that should be urgent priorities for frontier AI organizations today. These recommendations are critical to model weight security, most are feasible to achieve within about a year given prioritization, and they are not yet comprehensively implemented in frontier AI organizations:

- Develop a security plan for a comprehensive threat model focused on preventing unauthorized access and theft of the model's weights.
- Centralize all copies of weights to a limited number of access-controlled and monitored systems.
- Reduce the number of people authorized to access the weights.
- Harden interfaces for model access against weight exfiltration.
- Implement insider threat programs.
- Invest in defense-in-depth (multiple layers of security controls that provide redundancy in case some controls fail).
- Engage advanced third-party red-teaming that reasonably simulates relevant threat actors.
- Incorporate confidential computing to secure the weights during use and reduce the attack surface. (This measure is more challenging to implement than the others in this list but is backed by a strong consensus in industry.)

Securing future models against the most-capable threat actors will require stricter and more-advanced policies and systems. In particular, protecting models that are interacting with the internet against the most capable threat actors is currently not feasible. The development, implementation, and deployment of critical security measures needed to thwart such actors may take significant time (e.g., five years), and it is unclear whether such actions will take place without proactive solicitation. As a result, it would be wise to begin efforts toward these more-advanced measures soon. Examples of such efforts include

- physical bandwidth limitations between devices or networks containing weights and the outside world
- development of hardware to secure model weights while providing an interface for inference, analogous to hardware security modules in the cryptographic domain
- setting up secure, completely isolated networks for training, research, and other more advanced interactions with weights.

We emphasize that deciding on the appropriate security level for an organization or a specific AI model involves many considerations, including the capabilities of the models being secured, whether those capabilities pose potential threats to national security or competitiveness, and whether comparable capabilities are already available through other means. There is an ongoing, lively debate regarding the extent to which different models need to be secured (if at all). Our goal is to improve the ability to defend whichever frontier AI models are deemed worth securing at the desired security level by systemizing knowledge about which security postures achieve various desirable security outcomes, thus supporting informed decisionmaking in the private and public sectors.

Contents

Figures and Tables

Figures

Tables

CHAPTER 1

Introduction

The rapidly expanding capabilities of artificial intelligence (AI)/machine learning (ML) systems present both opportunities and risks. Recent advancements in AI hold the potential to significantly enhance labor productivity, human health, and other sectors.[1] However, this growth brings risks associated with AI misuse and unintended consequences of deployment, such as in the cybersecurity and biotechnology domains, which have been highlighted in such international statements as the Bletchley Declaration of the first AI Safety Summit and the United Nations (UN) General Assembly resolution, "Seizing the Opportunities of Safe, Secure and Trustworthy Artificial Intelligence Systems for Sustainable Development."[2] Addressing these challenges presents technical obstacles and uncertainties. There is already at least one known case (and arguably two) in which one of the most capable models of its time was irreversibly leaked.[3]

The motivation to secure frontier AI models includes not only protecting intellectual property but also potentially safeguarding national security. There has always been a commercial motivation to secure AI models. However, growing concerns that risks from future AI models may rise to national security significance introduce an additional motivation: the security and interests of the broader public.[4] As a result, discussions of how to secure *frontier* AI models—that is, models that match or exceed the capabilities of the most advanced AI models at the time of their development—are expanding beyond AI organizations to include stakeholders across industry, government, and the public.

Developing effective security measures for AI systems faces challenges due to the sophistication of potential threats, including high-priority operations by nation-states—a threat model often not referenced in discussions related to commercial companies but necessary to consider given the potential national security significance. Despite uncertainties in AI's development trajectory, immediate action is required for AI organizations to prepare for future security needs.

Communication about AI security strategies must extend beyond internal discussions at AI organizations. Organizations develop their own threat models, security strategies, and systems—and rightfully so: There is significant context and information that they cannot effectively or responsibly communicate

---

[1] Michael Chui, Eric Hazan, Roger Roberts, Alex Singla, Kate Smaje, Alex Sukharevsky, Lareina Yee, and Rodney Zemmel, *The Economic Potential of Generative AI: The Next Productivity Frontier*, McKinsey & Company, June 14, 2023; Chaitanya Adabala Viswa, Joachim Bleys, Eoin Leydon, Bhavik Shah, and Delphine Zurkiya, *Generative AI in the Pharmaceutical Industry: Moving from Hype to Reality*, McKinsey & Company, January 9, 2024.

[2] Government of the United Kingdom, "The Bletchley Declaration by Countries Attending the AI Safety Summit, 1–2 November 2023," webpage, November 1, 2023; UN General Assembly, 78th session, Seizing the Opportunities of Safe, Secure and Trustworthy Artificial Intelligence Systems for Sustainable Development, A/78/L.49, March 11, 2024.

[3] Carl Franzen, "Mistral CEO Confirms 'Leak' of New Open Source AI Model Nearing GPT-4 Performance," *GamesBeat*, January 31, 2024; James Vincent, "Meta's Powerful AI Language Model Has Leaked Online—What Happens Now?" *The Verge*, March 8, 2023.

[4] National Institute of Standards and Technology (NIST), "Biden-Harris Administration Announces New NIST Public Working Group on AI," press release, June 22, 2023a; Executive Order 14091, "Further Advancing Racial Equity and Support for Underserved Communities Through the Federal Government," Executive Office of the President, February 16, 2023.

externally. But an exclusively internally driven process also introduces challenges if an organization's responsibility extends beyond its own financial interests. An idiosyncratic view of one security team could have implications far wider than for the organization itself. In this context, policymakers need to be able to meaningfully engage with companies on their risk management strategies—whether through regulation or voluntary commitments. For example, multiple labs have published responsible scaling policies and preparedness frameworks.[5] These documents match the capabilities and risks of an AI model to a set of safety requirements (not limited to security) to which the organization is committed. To be effective, both in mitigating risk and demonstrating responsibility, there must be a shared understanding—not just within the organization but also across relevant stakeholders—of how an organization's security measures translate into actual security. That need applies not only to these voluntary frameworks but also to other forms of governance.

This report aims to promote more-robust AI security strategies by facilitating that shared understanding. To achieve this goal, we offer four important contributions, detailed in Figure 1.1.

FIGURE 1.1
Key Contributions of This Report

We identify approximately 38 meaningfully distinct attack vectors.

In most cases, an organization's vulnerability to just one of these vectors can compromise its security. We provide hundreds of real-world examples in which these attack vectors were deployed successfully, demonstrating that they are feasible and providing context on what such attacks look like in practice.

We explore a variety of potential attacker capabilities, from opportunistic (often financially driven) criminals to highly resourced nation-state operations.

This categorization of attacker capabilities allows organizations to identify sequential priorities depending on their current security infrastructure.

We estimate the feasibility of each attack vector being executed by different categories of attackers.

About a dozen attack vectors are likely infeasible for nonstate actors, but they are feasible for state actors, highlighting the need for significantly more-capable security systems to defend against state actors. Expert opinions vary significantly on the capabilities of state actors and how to defend against them.

We propose and define five security levels and recommend preliminary benchmark security systems that roughly achieve the security levels.

Each level is defined as being secure against attack vectors feasible for increasingly capable categories of malicious actors. The benchmarks can help to calibrate the trade-off between security investment and protection against different actors. The security levels are not meant to be used as a standard. Rather, they provide concrete suggestions for steps that frontier AI organizations can take at different stages of their continuous security enhancement strategy.

---

[5] Anthropic, "Anthropic's Responsible Scaling Policy," company announcement, September 18, 2023; OpenAI, *Preparedness Framework (Beta)*, December 18, 2023.

CHAPTER 2

Scope

The scope of this report is defined within the broad ecosystem of AI system security, encompassing components such as model weights, architectural design, training data, and operational infrastructure. **Given the size of this ecosystem, we have narrowed our focus to a critical subset: the learnable parameters of AI models, commonly referred to as model *weights*.** This includes both the weights and the biases learned during training, which are central to a model's ability to make predictions or decisions. We further focus exclusively on the risk from theft, copying, or mimicry of the weights of frontier AI models.

Our decision to concentrate on model weights stems from two key considerations:

- **Risk assessment:** Model weights uniquely represent the culmination of the different challenging prerequisites for training advanced models, including significant compute (i.e., the processing power and resources used to process data and run calculations, which is estimated at thousands of graphics processing unit [GPU]-years for GPT-4, with a reported $78 million and nearly $200 million in training costs for GPT-4 and Google's Gemini Ultra, respectively[1]) and training data (rumored to be more than 10TB for GPT-4),[2] algorithmic improvements and optimizations used during training, and more. Although the weights can be reproduced even if an attacker cannot exfiltrate them directly, reproducing them requires all of the above prerequisites. On the other hand, once the attacker has access to the weights of a model, abusing the model without restrictions or monitoring is eminently feasible. There are only two prerequisites. The first is the compute needed for inference, which is estimated to cost less than $0.005 per thousand tokens,[3] or roughly $0.0065 per word. The second is the model architecture—the preexisting structure of the model before training begins—which is harder to secure than the weights themselves (see below) and might be inferred from the structure of the weights.
- **Feasibility:** Securing model weights is technically challenging, yet it offers a more tractable point of intervention than other components, such as architecture or training data. The model architecture is a much smaller piece of information and thus is much easier to exfiltrate, either through a network or in the mind of an engineer. There are also many more people who need to know the architecture details: researchers identifying model improvement, engineers optimizing model efficiency, and more. While model weights are frequently used, most use cases do not require flexible read access to the full weights—the weights can be more easily protected by copy-resistant interfaces (see the "Permitted Interfaces" subcategory for Security Level 3 in Chapter 6 and Appendix B). Training data are often scraped from public sources or purchased from commercial aggregators and therefore less under the exclusive control of the AI organization.

---

[1] Artificial Intelligence Index, *Artificial Intelligence Index Report 2024*, Stanford University, 2024.

[2] Dylan Patel and Gerald Wong, "GPT-4 Architecture, Infrastructure, Training Dataset, Costs, Vision, MoE," *SemiAnalysis*, July 10, 2023.

[3] Tokens are portions of words processed by large language models (LLMs). On average, about every three words are translated into four tokens before being fed into the model.

Our analysis focuses on foundation models, and specifically large language models (LLMs) and similar multimodal models. We make the following key **technical assumptions** about such models:

- **Size:** Frontier models are large (reaching terabytes of required weight storage[4]) and expected to grow substantially in the future,[5] making their unauthorized duplication or theft easier to monitor or prevent.
- **Availability:** Common use cases for these models require high availability online, typically through an inference application programming interface (API), which introduces challenging constraints on the ways such models can be isolated (at least in the commercial context).

We expect most findings and recommendations to generalize to other frontier AI models that also have these properties.

We exclude models whose weights are not deemed critical to secure. One reason a model's weights may be deemed critical to secure is that its capabilities pose a risk to public safety. Assessment of whether a model poses a large-scale societal risk is a nascent field, but in the future different security measures proposed in this report may be applied depending on the assessed risk of each model (or applied by default to potentially risky models that have not yet been assessed). Once a model has been made publicly available (which is often referred to as "open-sourcing" it), there is no longer value in securing specific copies of it. The decision of whether to "open-source" future models should be informed by whether their risks justify controlling access to them.

Although we focus on model weights, we acknowledge the importance of other aspects of AI security. Securing the confidentiality of other components, such as the model architecture, training data, and source code, plays a crucial role in an AI system's overall security posture but is beyond this report's purview. Similarly, protecting the model's integrity and availability, guarding against the misuse of legitimate APIs, and planning for harm mitigation in case the model is exfiltrated also play important roles.

We focus on measures for the AI organizations themselves to implement to improve their own security, though we aim to make the discussion accessible to other stakeholders who may engage with the organizations on their security. Governments and the broader research and development (R&D) community can also take further action to support the security of frontier AI models, but such efforts are beyond the scope of this report.

---

[4] The storage size of GPT-4's model parameters is estimated to be in the terabytes (Latent Space, "Commoditizing the Petaflop—with George Hotz of the Tiny Corp," webpage and video, June 20, 2023). The ratio between parameter count and storage size can be estimated based on open-source models.

[5] Epoch AI, "Announcing Epoch AI's Updated Parameter, Compute and Data Trends Database," October 23, 2023.

CHAPTER 3

Methodology

Our methodology was designed to aggregate and synthesize existing knowledge on risks associated with unauthorized access to AI model weights and provide recommendations for mitigating such risks. In this chapter, we first discuss the information sources we use throughout the report and then describe the process we used to develop the report using these sources.

Sources

Interviews

We conducted interviews with 32 experts: 6 national security government personnel specializing in information security, 6 prominent information security industry experts, 8 senior information security staff from frontier AI companies, 6 other senior staff from frontier AI companies, 4 independent AI experts with prior experience at frontier AI organizations, and 2 insider threat experts. These interviews are a primary source of information in this report (particularly for our benchmarks and recommendations). These interviews naturally yielded a spectrum of viewpoints, with occasional significant disagreements. We have done our best to synthesize our interviewees' views in ways that reflect our best understanding of their substance, and we have indicated in our discussion some of the important issues on which our experts differed most strongly. However, we emphasize that the fact that a recommendation or conclusion appears in the report does not imply that all consulted experts supported it.

Our interview protocol appears in Appendix D.

Literature Review

Additionally, we analyzed a variety of written sources from the academic literature, commercial security reports, official government documents, media reports, and other online sources—primarily to empirically back our statements, put claims in context, and gain inspiration for how to structure information in the report. Existing frameworks and taxonomies we drew on include the NIST Cybersecurity Framework,[1] the MITRE ATT&CK® framework,[2] and such AI-specific sources as the MITRE Adversarial Threat Landscape for Artificial-Intelligence Systems (ATLAS),[3] the NIST Adversarial Machine Learning: A Taxonomy and Terminology of Attacks and Mitigations,[4] the Berryville Institute of Machine Learning (BIML) Interactive

[1] NIST, *The NIST Cybersecurity Framework 2.0*, NIST CSWP 29 (Initial Public Draft), August 8, 2023c.

[2] This includes MITRE ATT&CK (MITRE, "ATT&CK," webpage, undated-b) and the MITRE ATT&CK Enterprise Matrix (MITRE, "ATT&CK—Enterprise Matrix," webpage, undated-c).

[3] MITRE, "ATLAS," webpage, undated-a.

[4] Alina Oprea and Apostol Vassile, *Adversarial Machine Learning: A Taxonomy and Terminology of Attacks and Mitigations*, National Institute of Standards and Technology, NIST AI 100-2 E2023 (Initial Public Draft), March 8, 2023.

Machine Learning Risk Framework,[5] and Microsoft's "Failure Modes in Machine Learning."[6] Additional sources are cited throughout the report.

Process

Step 1: Development of Operational Capacity Categories (Chapter 4)

In consultation with experts, we developed a set of five *operational capacity* categories to roughly classify offensive cyber operations by their resources and capabilities, offering a common vocabulary for subsequent discussions in the report. These categories represent thresholds along a continuum of capabilities, not step changes between completely distinct groups.

Step 2: Identification of Attack Vectors (Chapter 5)

We compiled a list of potential attack vectors that could be used to access and exfiltrate model weights, based primarily on the literature and suggestions from our interviewed experts. For each suggested attack vector, we searched for evidence of its successful use in real-world attacks on digital systems. We included an attack vector in the final list if we could find evidence of successful real-world execution, if a majority of experts who commented on the attack vector deemed it to be nascent but likely (e.g., for several of the AI-specific attack vectors), or if multiple experts testified that real-world executions exist but evidence was not publicly available (e.g., for a small number of advanced attack vectors used by some nation-states). In the few cases for which we could not provide direct evidence of real-world execution, we provided similar or analogous examples that aim to shed light on the feasibility of the approach.

The MITRE ATT&CK framework is a well-known framework for adversary tactics and techniques that has (directly and indirectly) influenced our structuring of attack vectors. The MITRE ATLAS framework plays a similar role in the AI-specific context. However, the goals of our taxonomy of attack vectors differed sufficiently from these frameworks that we decided not to use them to structure our discussion. Departing from existing frameworks allows us to focus more deeply on protecting AI model weights from theft while avoiding unnecessary additional terminology and structure that could reduce clarity for readers outside the cyber security industry.

The full list of attack vectors, along with the real-world execution examples and additional commentary, is available in Appendix A.

Step 3: Capability Assessment for Attack Vectors (Chapter 5)

We estimated the capability of members of each operational capacity category to execute each attack vector. This was done in consultation with the interviewed experts through the following process. Experts were initially asked to provide estimates of feasibility based on the scoring system described in Chapter 5; in following iterations, we presented experts with the tentative estimates produced and elicited feedback on the accuracy of the estimates and any suggested changes. Finally, we held in-person workshops that allowed the experts to directly interact with each other and provide additional feedback to the scores based on their discussions. Generally, most experts provided feedback for a minority of attack vectors, but they were encouraged to prioritize commenting on estimates with which they disagreed or thought were mistaken.

---

5 BIML, "BIML Interactive Machine Learning Risk Framework," webpage, undated.

6 Microsoft, "Failure Modes in Machine Learning," November 2, 2022.

The feasibility scores presented in Table 5.2: Capability Assessment represent a rough (rounded) average of the expert opinions we received. Some experts provided qualitative feedback, ranges, and other informative but not fully quantitative feedback; as a consequence, the scores in the table are not mathematical averages but rather a conceptual "center of mass" of shared expert opinion. We disregarded claims advocated by experts for a feasibility score that is directly disputed by existing evidence (e.g., a claim that a certain attack vector is infeasible for a certain category despite the fact that multiple cases of actors within that group executing the attack vector exist).

Step 4: Development of Security Levels and Their Benchmarks (Chapter 6)

Finally, we defined the five *security levels* as the level of security required to secure a system against each of the five operational capacity categories. We collected the potential security measures to be included from governmental security guidance, security reports, the academic literature, and the interviewed experts. We then assigned security measures to specific security level benchmarks. This step was informed by the assessed feasibility of the different attack vectors for the different operational capacity categories: Security measures that aim to secure against a specific attack vector are in the lowest security level that has a corresponding operational capacity category that includes operations for which the attack vector is feasible (determined by a feasibility score above 1).

Other considerations taken into account in forming the security levels include the interactions between security measures, the need for defense-in-depth (i.e., multiple layers of security controls that provide redundancy in case some controls fail), and other vector-independent factors. To incorporate these considerations, we developed the security level benchmarks through iterative consultations with experts. These benchmarks are thus informed by additional expert insights rather than directly implied by the previous steps. We endeavored to incorporate all expert feedback we received and to identify reasonable compromises when expert feedback conflicted. Because of the complex, qualitative, and multidimensional nature of the security level benchmarks, this was done through general engagement with expert feedback and adaptation rather than a strict formulaic process.

The NIST Cybersecurity Framework is a foundational framework for managing cybersecurity risk and has influenced many cybersecurity professionals, including many of our interviewed experts. However, despite multiple structural similarities, we decided not to construct our taxonomy of security measures according to categories of the NIST Cybersecurity Framework because we found the ability to structure according to AI-specific categories more important to clarity. For those familiar with the NIST framework, Appendix B maps our categories to the NIST Cybersecurity Framework.

CHAPTER 4

Defining Operational Capacity Categories

We define five *operational capacity* categories based on the resources and capabilities available to the operation (see Figure 4.1). Recognizing the complexity of attackers, which vary in funding, expertise, workforce size, preexisting organization access, and infrastructure,[1] we simplify these factors into a single scale. Our categories are monotonic: By definition, each category includes the capacities of all preceding ones. Introducing some concept of increasingly capable threat actors is necessary as AI organizations and policymakers seek to map AI model capabilities to incremental security requirements. See additional discussion of the need for incremental threat actor categories, and more importantly incremental security levels, in Chapter 6.

Despite the limitations of such simplification, this approach facilitates a rough comparison between threat actors. By explicitly stating the assessed capabilities of different categories of operational capacity, we increase readers' ability to adjust their needs and requirements to the types of threats they foresee or are concerned about. When considering their own threat models, readers can identify which category "best fits" the details of the specific operation and, if necessary, adapt their security posture where their considered threats differ from the assessed capabilities for the category (see also Table 5.2: Capability Assessment in Chapter 5).

In Figure 4.1, the title of each operational capacity category should be seen as an intuitive example of what actors might fall into that category—the actual definition of each category follows the title and supersedes it.

[1] Some of these aspects are qualitatively annotated and described in OASIS, "STIX™ Version 2.0. Part 1: STIX Core Concepts—Committee Specification 01," webpage, July 19, 2017.

FIGURE 4.1
Operational Capacity Definitions

OC1 Amateur attempts

Operations roughly less capable than or comparable to a single individual with some limited professional expertise in information security spending several days with a total budget of up to $1,000 on the specific operation, and no preexisting infrastructure or access to the organization.

This includes the operations of many **hobbyist hackers**, as well as more experienced hackers who implement completely untargeted **"spray and pray" attacks**.

OC2 Professional opportunistic efforts

Operations roughly less capable than or comparable to a single individual who is broadly capable in information security spending several weeks with a total budget of up to $10,000 on the specific operation, with preexisting personal cyber infrastructure but no preexisting access to the organization.

This includes the operations of many **individual professional hackers**, as well as **capable hacker groups when executing untargeted or lower-priority attacks**.

OC3 Cybercrime syndicates and insider threats

Operations roughly less capable than or comparable to ten individuals who are experienced professionals in information security spending several months with a total budget of up to $1 million on the specific operation, with major preexisting cyberattack infrastructure but no preexisting access to the organization. Also included in this category are attempts by insider threats within the organization, who will have significantly less resources and expertise than the previous operations described as part of this category but significant access to sensitive organization resources (e.g., a senior member of the organization's research team).

This includes the operations of many **world-renowned criminal hacker groups, well-resourced terrorist organizations, disgruntled employees**, and **industrial espionage organizations.**[a]

OC4 Standard operations by leading cyber-capable institutions

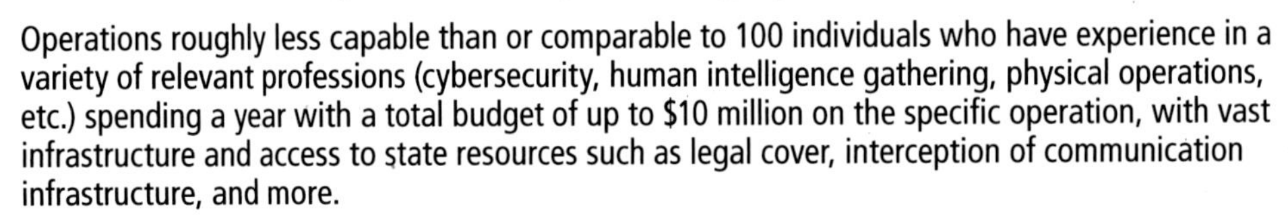

Operations roughly less capable than or comparable to 100 individuals who have experience in a variety of relevant professions (cybersecurity, human intelligence gathering, physical operations, etc.) spending a year with a total budget of up to $10 million on the specific operation, with vast infrastructure and access to state resources such as legal cover, interception of communication infrastructure, and more.

This includes the operations of many of the world's leading **state-sponsored groups** and many **foreign intelligence agencies** across the world. The top cyber-capable nations globally can execute such operations more than **100 times per year**.

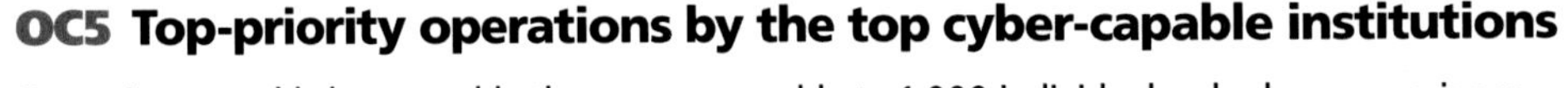

OC5 Top-priority operations by the top cyber-capable institutions

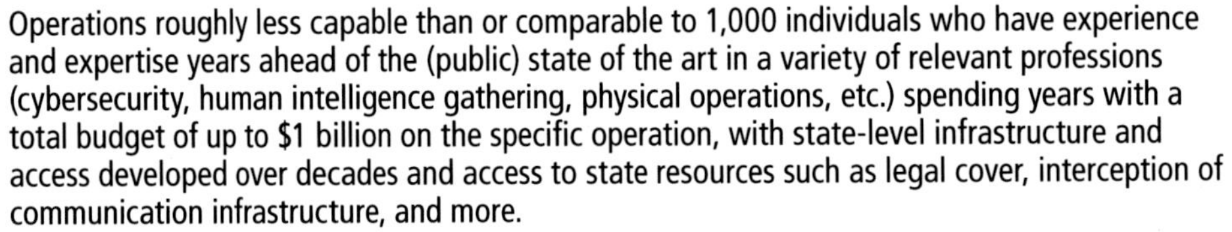

Operations roughly less capable than or comparable to 1,000 individuals who have experience and expertise years ahead of the (public) state of the art in a variety of relevant professions (cybersecurity, human intelligence gathering, physical operations, etc.) spending years with a total budget of up to $1 billion on the specific operation, with state-level infrastructure and access developed over decades and access to state resources such as legal cover, interception of communication infrastructure, and more.

This includes the handful of operations most prioritized by the world's **most capable nation-states.**

[a] The set of actors within OC3 is more diverse than in other categories—most notably in the inclusion of both insider threats and external cyber organizations. We group the OC3 actors together because the level of investment required to robustly defend against them is comparable, despite the specific measures required being partially but not fully overlapping.

CHAPTER 5

Analysis of Potential Attack Vectors

Having established five categories of adversarial operational capacity, we now enumerate various strategies that attackers could deploy to steal model weights—38 attack vectors in total, divided into 9 categories. We then estimate the feasibility of each attack vector by operational capacity category, which will inform which attack vectors need to be secured against when protecting against different threat actors.

The attack vectors can be considered a subset of a broader *cyber kill chain*: the set of all actions taken by attackers to conduct offensive cyber operations. The attack vectors focus on those components that directly advance an attacker in breaching system defenses to reach, gain access to, and exfiltrate model weights. This narrower view allows identifying and effectively communicating many diverse security needs in a way that is accessible to audiences not familiar with all technical aspects of offensive cyber operations. We touch on other aspects of the cyber kill chain not captured in this taxonomy later in this chapter and in subsequent chapters. Importantly, the attack vectors are not limited to actions taken in the cyber domain.[1]

These attack vectors are not merely theoretical. Empirical evidence (described in Appendix A) suggests that they pose concrete risks. The vast majority of them have been deployed in real-world environments.

Most security experts will find a majority of the attack vectors we highlight familiar; we include a discussion and examples of these more "mundane" attack vectors as an introduction for readers less versed in information security. However, even some senior security experts are unfamiliar with certain offensive tools or presume them to be infeasible—especially those in the arsenal of highly resourced state actors.

We reiterate that the attack vectors we present are focused on the theft of model weights; other vectors that are relevant for the security of AI systems exist. Because these vectors are less likely to be part of a weight-theft attack chain, we do not include them (e.g., training data poisoning, prompt injection unrelated to code execution).

Landscape of Attack Vectors

Table 5.1 summarizes the nine categories of attack vectors we assess. The definitions of each attack vector, detailed descriptions, comments on their typical characteristics and severity, and multiple real-world examples appear in Appendix A.

[1] Lockheed Martin, "Cyber Kill Chain," webpage, undated.

TABLE 5.1
Summary of Attack Vectors

Attack Category	Attack Vector
Running Unauthorized Code	• Exploiting vulnerabilities for which a patch exists (attacking non-updated software) • Exploiting reported but not (fully) patched vulnerabilities • Finding and exploiting individual zero-days[a] • Direct access to zero-days at scale
Compromising Existing Credentials	• Social engineering • Password brute-forcing and cracking • Exploitation of exposed credentials • Expanding illegitimate access (e.g., escalating privileges)
Undermining the Access Control System Itself	• Encryption/authentication vulnerabilities (in the access control system) • Intentional backdoors in algorithms, protocols, or products (in the access control system) • Code vulnerabilities (in the access control system) • Access to secret material undermining a protocol
Bypassing Primary Security System Altogether	• Incorrect configuration or security policy implementation • Additional (less secure) copies of sensitive data • Alternative (less secure) authentication or access schemes
AI-Specific Attack Vectors	• Discovering existing vulnerabilities in the ML stack • Intentional ML supply chain compromise • Prompt-triggered code execution • Model extraction • Model distillation
Nontrivial Access to Data or Networks	• Digital access to air-gapped networks • Side-channel attacks (including through leaked emanations; i.e., TEMPEST attacks) • Eavesdropping and wiretaps
Unauthorized Physical Access to Systems	• Direct physical access to sensitive systems • Malicious placement of portable devices • Physical access to devices in other locations • Evasion of physical access control systems • Penetration of physical hardware security • Armed break-in • Military takeover
Supply Chain Attacks	• Services and equipment the organization uses • Code and infrastructure incorporated into the codebase • Vendors with access to information
Human Intelligence	• Bribes and cooperation • Extortion • Candidate placement • Organizational leverage attacks • Organizationally approved access

[a] *Zero-days* are vulnerabilities that have not yet been identified or mitigated by the vendor or the broad cybersecurity community (i.e., there have been at most "zero days" since the vendor discovered or mitigated the vulnerability).

Key Conclusions About the Attack Vector Landscape

The diversity of attack vectors is large, so defenses need to be varied and comprehensive. Achieving strong security against a specific category of attack does not protect an organization from others.

The diversity of attack vectors is large, so defenses need to be varied and comprehensive.

Many attack vectors are widely accessible, some more so than even many information security experts are currently aware of (e.g., a $180 USB cable can provide full remote control of a device).[2]

Even organizations that have invested heavily in security have suffered severe breaches of sensitive systems and information. Famous examples include multiple U.S. government intelligence agencies,[3] the Iranian nuclear program,[4] U.S. nuclear power plants,[5] Google,[6] Microsoft,[7] and others.

The U.S. Annual Threat Assessment Report estimates that multiple foreign agencies are able to penetrate and disrupt varied types of U.S. critical infrastructure, so the requirements and resources made available to critical infrastructure sectors are not sufficient to consistently lead to security against foreign nations.[8]

The scale of operations and investments by highly resourced state actors can be inferred from several data points:

- A 2013 report in the *Washington Post* indicated the massive scale of such operations.[9] In 2011, U.S. spy agencies executed 231 cyber operations and had plans to place millions of "implants" (both physical and digital).[10]
- In 2020, the U.S. Department of Defense allocated a budget of at least $3.7 billion to cyber operations,[11] with an additional $2.6 billion for training.
- Federal Bureau of Investigation (FBI) Director Christopher Wray implied that China has a workforce of more than 175,000 hackers.[12]

---

[2] Hak5, "O.MG Cable," webpage, undated-a.

[3] Scott Shane, Nicole Perlroth, and David E. Sanger, "Security Breach and Spilled Secrets Have Shaken the N.S.A. to Its Core," *New York Times*, November 12, 2017; Greg Miller and Ellen Nakashima, "WikiLeaks Says It Has Obtained Trove of CIA Hacking Tools," *Washington Post*, March 7, 2017.

[4] Catherine A. Theohary, *Iranian Offensive Cyber Attack Capabilities*, Congressional Research Service, IF11406, January 13, 2020.

[5] Salih Bıçakcı, *Introduction to Cyber Security for Nuclear Facilities*, Centre for Economics and Foreign Policy Studies, 2015.

[6] Kim Zetter, "'Google' Hackers Had Ability to Alter Source Code," *Wired*, March 3, 2010.

[7] Mitchell Clark, Richard Lawler, and Jay Peters, "Microsoft Confirms Lapsus$ Hackers Stole Source Code via 'Limited' Access," *The Verge*, March 22, 2022.

[8] National Intelligence Council, *Annual Threat Assessment of the U.S. Intelligence Community*, Office of the Director of National Intelligence, February 6, 2023.

[9] Wilson Andrews and Todd Lindeman, "$52.6 Billion: The Black Budget," *Washington Post*, August 29, 2013.

[10] Barton Gellman and Ellen Nakashima, "U.S. Spy Agencies Mounted 231 Offensive Cyber-Operations in 2011, Documents Show," *Washington Post*, August 30, 2013.

[11] Mark Pomerleau, "What's in the $9.6B Cyber Budget Request?" *C4ISRNET*, March 14, 2019.

[12] Christopher Wray, "Director's Opening Statement to the House Committee on Appropriations Subcommittee on Commerce, Justice, Science, and Related Agencies," testimony, Federal Bureau of Investigation, April 27, 2023.

- North Korea's investments in cyber operations allegedly range "between 10% to 20% of the regime's military budget," or around $400 million to $800 million.[13]

Highly resourced operations by state actors (equivalent to our OC5 category) are the apex of cyber operations, characterized by their complexity, associated risks, and access to dedicated teams that can tailor research and operations to each targeted organization and the immense resources required. Highly resourced operations by state actors are executed infrequently, even by the most advanced states, but their financial and technical scale, combined with the secrecy surrounding these operations, make them challenging to defend against.

Capability Assessment for Attack Vectors

While the execution of some attack vectors is ubiquitous, others are deployed only rarely, and their feasibility is a matter of some debate. The question of how frequently different attack vectors are deployed and by whom draws an even larger diversity of views.

In this section, we present aggregated information about the estimated feasibility of the different attack vectors. This information helps inform requirements for the security level benchmarks presented in Chapter 6 but can also assist organizations in roughly calibrating to the assessments of other experts when developing their own threat models, fine-tuned to their specific circumstances and private information.

Table 5.2 presents assessments of the feasibility of each attack vector from different types of actors, on a scale of 1 (low feasibility) to 5 (high feasibility). Additional detail on how we developed the capability scores—and dealt with uncertainty about feasibility—is presented in Box 5.1.

[13] Dave Lee and Nick Kwek, "North Korean Hackers 'Could Kill,' Warns Key Defector," BBC News, May 29, 2015.

TABLE 5.2
Capability Assessment

Attack Category	Attack Vector	OC1—Amateur Attempts	OC2—Professional Opportunistic Efforts	OC3—Cybercrime Syndicates or Insider Threats	OC4—Standard Operations by Major Cyber-Capable Institutions	OC5—Top-Priority Operations by the Top Cyber-Capable Institutions
Running Unauthorized Code	Exploiting vulnerabilities for which a patch exists (attacking non-updated software)	3	4	4	5	5
	Exploiting reported but not (fully) patched vulnerabilities	2	2	3	4	5
	Finding and exploiting individual zero-days	1	2	4	5	5
	Direct access to zero-days at scale	1	1	2	4	5
Compromising Existing Credentials	Social engineering*	3	4	5	5	5
	Password brute-forcing and cracking*	2	2	3	3	4
	Exploitation of exposed credentials	3	3	4	5	5
	Expanding illegitimate access (e.g., escalating privileges)	2	3	5	5	5
Undermining the Access Control System Itself	Encryption/authentication vulnerabilities (in the access control system)	1	1	2	3	3
	Intentional backdoors in algorithms, protocols, or products (in the access control system)	1	1	1	2	2
	Code vulnerabilities (in the access control system)	1	1	2	3	4
	Access to secret material undermining a protocol	1	1	2	2	3
Bypassing Primary Security System Altogether	Incorrect configuration or security policy implementation	2	2	3	3	3
	Additional (less secure) copies of sensitive data	2	2	3	5	5
	Alternative (less secure) authentication or access schemes	1	1	2	3	4

Table 5.2—Continued

Attack Category	Attack Vector	OC1—Amateur Attempts	OC2—Professional Opportunistic Efforts	OC3—Cybercrime Syndicates or Insider Threats	OC4—Standard Operations by Major Cyber-Capable Institutions	OC5—Top-Priority Operations by the Top Cyber-Capable Institutions
AI-Specific Attack Vectors	Discovering existing vulnerabilities in the ML stack*	1	2	4	4	5
	Intentional ML supply chain compromise*	1	2	4	5	5
	Prompt-triggered code execution*	2	3	4	4	4
	Model extraction*	1	1	1	2	3
	Model distillation*	1	1	2	2	3
Nontrivial Access to Data or Networks	Digital access to air-gapped networks	1	1	2	2	4
	Side-channel attacks (including through leaked emanations; i.e., TEMPEST attacks)	1	1	1	2	3
	Eavesdropping and wiretaps	1	1	2	4	5
Unauthorized Physical Access to Systems	Direct physical access to sensitive systems	1	1	3	3	4
	Malicious placement of portable devices	2	2	4	4	4
	Physical access to devices in other locations	1	1	3	4	5
	Evasion of physical access control systems	1	1	1	3	5
	Penetration of physical hardware security	1	1	1	2	3
	Armed break-in	1	1	1	2	4
	Military takeover	1	1	1	1	2
Supply Chain Attacks	Services and equipment the organization uses	1	2	4	5	5
	Code and infrastructure incorporated into the codebase	1	2	4	5	5
	Vendors with access to information	1	1	2	3	4

Table 5.2—Continued

Attack Category	Attack Vector	OC1—Amateur Attempts	OC2—Professional Opportunistic Efforts	OC3—Cybercrime Syndicates or Insider Threats	OC4—Standard Operations by Major Cyber-Capable Institutions	OC5—Top-Priority Operations by the Top Cyber-Capable Institutions
Human Intelligence	Bribes and cooperation	1	1	4	4	5
	Extortion	1	1	4	4	5
	Candidate placement	1	1	1	3	5
	Organizational leverage attacks	1	1	1	3	5
	Organizationally approved access	1	1	2	3	5

NOTE: An asterisk (*) indicates that we believe that the nature of the attack could change very rapidly. The table cells provide a score reflecting the likelihood that a single arbitrary actor from a certain operational capacity category will successfully execute an attack against a single arbitrary target. A score of 1 represents up to a 20 percent chance of success, 2 represents a 20–40 percent chance of success, 3 represents a 40–60 percent chance of success, 4 represents a 60–80 percent chance of success, and 5 represents more than 80 percent chance of success.

BOX 5.1

Interpreting the Capability Scores

Questions around the feasibility of capabilities can be abstruse. To ensure that the experts we interviewed were answering the same question, and to assist readers in correctly interpreting the scores in Table 5.2, we describe here how the probability of success that the scores estimate is defined. These definitions were also provided to the experts during discussions and feedback iterations.

The capability scores aim to (roughly) represent the likelihood that a single operation from the relevant operational capacity category can successfully execute the attack in a real-world scenario. Note that if many attempts of a specific operational capacity category are likely to occur, then the aggregate likelihood that one of them will succeed will be higher.

The victim of the perpetrated attack is assumed to be an average tech company with only ubiquitously available security mechanisms (equivalent to the benchmark for SL1 described in Appendix B).

However, if the attack vector focuses on overcoming a specific defense (e.g., penetration of physical hardware security), then this defense is assumed to exist in the target system.

An attack is considered successful if its execution directly contributes to the attacker's ability to steal model weights, even if the successful attack does not on its own provide full access to the weights—i.e., if the attack is only one part of a longer cyber kill chain. For example, if an attacker is likely to find a code vulnerability that allows them to execute code on a random company device with no useful permissions, this is not counted as a successfully executed attack. However, if that code vulnerability enables an attacker to advance their reach toward the weights (even if other attack vectors need to be used to fully exfiltrate the weights), then it is counted as a successfully executed attack.

The estimate of likelihood aims to represent the intersection of the vulnerability existing in the target system, the attacker being able to exploit the vulnerability (e.g., due to existing infrastructure, the ability to perform other attacks that are prerequisites for this one, etc.), and the attacker being willing to move forward with executing it (given the costs, risk, and capacity requirements involved) when necessary to steal model weights.

Interpreting the Capability Scores—Continued

All operational capacity categories are diverse. OC3, in particular, includes a large variety of actors (hackers, disgruntled employees, terrorists, etc.; see the discussion on the diversity of capabilities within this category in the "Bottom Lines" section in the discussion of SL3 in Chapter 6). The scores in Table 5.2 represent the actor most capable of performing the attack within that category.

The score estimates current feasibility. The feasibility of some attack vectors might change substantially over time.

Notable Areas of Disagreement and Consensus

As indicated earlier, among the subject-matter experts with whom we spoke, there was a diversity of opinions around the existence of many attack vectors and the feasibility of implementing them. Table 5.2 aims to capture a conceptual average of expert opinion; however, it does not imply that all experts agreed with the scores. The following aspects of the attack vectors generated more or less variation in opinions:

- Opinions about threats that fall into the realm of digital cybersecurity (i.e., the categories Running Unauthorized Code, Compromising Existing Credentials, Undermining the Access Control System Itself, and Bypassing Primary Security System Altogether) showed relatively little variation. On the other hand, other attack vectors that are outside the digital cybersecurity realm, that are not a common topic of academic research, or whose deployment is limited to a smaller set of threat actors (including the categories Nontrivial Access to Data or Networks, Unauthorized Physical Access to Systems, Supply Chain Attacks, and Human Intelligence) received very varied responses. In particular, side-channel attacks and direct access to zero-days at scale drew many differing opinions.
- There was significant uncertainty around the feasibility of the AI-specific attack vectors, both currently and in the future. However, that there was uncertainty was a point of consensus rather than disagreement: That is, most experts agreed that the uncertainty surrounding the feasibility of this attack vector was substantial. This uncertainty should be a major concern for stakeholders seeking security assurances, and therefore more research here is warranted.
- Across all attack vectors, there was significantly greater convergence in views on their feasibility for less-capable actors (OC1–OC3) than for more-capable actors (OC4–OC5).
- Concern around the feasibility of threats from organization insiders (which span many of the categories and are not limited to the Human Intelligence category) was an emerging point of consensus. Experts generally agreed that this is a significant source of risk and that significant efforts should be dedicated to mitigating it.

Concluding Remarks for the Attack Vectors

The analysis in this chapter provides a foundational overview of the diversity of attack vectors that organizations may face in securing their model weights, evidence of the feasibility of such attacks, and the operational capacity categories that may be positioned to execute the attacks. While not exhaustive, these illustrative attack vectors offer a practical starting point for modeling threats and prioritizing defenses.

For example, eight of the attack vectors are likely infeasible for OC1–OC3 (with a score of 1 in Table 5.2: Capability Assessment) but are likely feasible for OC4 or OC5 (with a score greater than 1 in the table), implying specific security measures needed to secure against these categories but not needed in systems meant to be secure only against less capable adversaries.

For multiple reasons, it is prudent to recognize the plausibility of current assessments underestimating the threat:

- We assume that other attack vectors exist that are as yet unknown to security experts, particularly ones concerning advanced persistent threats (APTs), such as state actors.
- Novel attack vectors and conceptual approaches are likely to evolve over time, as are novel insights and infrastructure that make existing attacks more accessible.
- Publicly known examples of attacks are only a subset of attacks actually taking place, especially when it comes to more-advanced operations. Most APTs persist for years before discovery.[14] Many national security experts with whom we spoke mentioned that the vast majority of highly resourced state actor attacks they are aware of were never publicly revealed. This means that a purely empirical analysis based on detected operations would systematically underestimate the feasibility and frequency of advanced attack vectors.
- Accordingly, one should expect capable actors to have access not only to well-established attack vectors but also to unknown approaches. In Appendix A, we share many examples of state actors developing such conceptually novel attacks years or decades before they were discovered by others.
- A common method for estimating the feasibility of a specific attack by different actors (or assessing the security of a system) is to estimate the cost of the attack (or the cost to infiltrate the system)—for example, one might estimate that "it costs $500,000 to buy or develop a Chrome remote code execution vulnerability." However, for APTs and especially for state actors with significant infrastructure, the marginal cost of executing an attack given preexisting investments is much lower than the total end-to-end cost of the attack (including capability development). Such actors routinely develop and maintain such capabilities, which can be used multiple times.
- Most of the analysis regarding state actors assumes an attempt by a foreign nation to surreptitiously exfiltrate model weights. Organizations based in countries where intelligence services have a history of using the power of the state to conduct cyber or physical operations on their own soil will have a significantly harder time securing their weights than our discussion implies.

This attack vector analysis offers an instructive slice of the larger security landscape. Other important aspects not captured by assessing the feasibility of attack vectors in isolation include capabilities related to other components of the full cyber kill chain and the ability to creatively string these attack vectors together.[15] Although the explicit analysis of these aspects is beyond the scope of this chapter of the report, they influence the security levels (Chapter 6) indirectly through the assessments and recommendations of the interviewed experts. The analysis of the threat landscape provided in this chapter can also serve as a practical resource for organizations taking steps now to calibrate their threat models in the face of an uncertain threat landscape.

---

[14] Lillian Ablon and Andy Bogart, *Zero Days, Thousands of Nights: The Life and Times of Zero-Day Vulnerabilities and Their Exploit*, RAND Corporation, RR-1751-RC, 2017.

[15] Lockheed Martin, undated.

CHAPTER 6

Security Levels

What Are the Security Levels?

To facilitate more-nuanced discourse on the security needs of different AI systems (depending on their capabilities or other safety concerns), we propose five security levels (SLs)—SL1 to SL5—broadly defined as the level of security a system requires to thwart increasingly capable operations. For instance, SL2 is defined as a system that is protected against most professional opportunistic attempts; SL5 is a system that is protected from top-priority operations by the top cyber-capable institutions. See Figure 6.1 for the full definitions of the security levels.

Frameworks for assessing the security posture of systems must balance generalizability and adaptability with specificity and concreteness. In recent years, general frameworks for securing digital systems (such as the NIST Cybersecurity Framework[1]) have trended away from using predefined security levels and toward guiding each organization to define its own security requirements. This approach has made it easier for organizations to tailor their security to their needs. However, without other means of calibrating threat models, security measures, and expected security outcomes, different organizations can reach wildly different conceptions of what is needed to achieve robust security against threat actors of interest. This is reasonable and perhaps beneficial in cases where organizations are primarily responsible for their own economic well-being, but it presents a challenge in cases where the security of those organizations has broad societal implications.

On the other hand, cybersecurity frameworks that are scoped for a more narrow system type, goal, or industry (such as the International Society of Automation/International Electrotechnical Commission [ISA/IEC] 62443 Series of Standards,[2] the Federal Information Processing Standard [FIPS] 140-3 Standard,[3] and Supply-chain Levels for Software Artifacts [SLSA][4]) may prioritize clarity and concreteness over flexibility, because the systems they consider have more in common. Our security level benchmarks fall closer to the prescriptive approach: By designating security goals for each level and associating them with suggested security measures (detailed below), we provide clarity and specificity that is difficult to achieve in more-flexible frameworks.

Both approaches have their limitations and risks. An overly flexible framework may fail to concretely inform adherents who follow the instructions but whose implementation remains insecure. An overly prescriptive framework may redirect attention from the specific security needs of an organization toward a predefined set of requirements.

---

1 NIST, "Cybersecurity Framework," webpage, undated.

2 International Society of Automation, "The World's Only Consensus-Based Automation and Control Systems Cybersecurity Standards," ISA/IEC 62443 Series of Standards, undated.

3 NIST, *Security Requirements for Cryptographic Modules*, Federal Information Processing Standards Publication 140-3, March 22, 2019.

4 Supply-chain Levels for Software Artifacts, homepage, undated.

FIGURE 6.1
The Five Security Levels

SL1 A system that can likely thwart amateur attempts (OC1).

This includes the operations of many hobbyist hackers, as well as more experienced hackers who implement completely untargeted "spray and pray" attacks.

SL2 A system that can likely thwart most professional opportunistic efforts by attackers that execute moderate-effort or nontargeted attacks (OC2).

This includes the operations of many professional individual hackers, as well as capable hacker groups when executing untargeted or lower-priority attacks.

SL3 A system that can likely thwart cybercrime syndicates or insider threats (OC3).

This includes the operations of many world-renowned criminal hacker groups, well-resourced terrorist organizations, disgruntled employees, and industrial espionage organizations.

SL4 A system that can likely thwart most standard operations by leading cyber-capable institutions (OC4).

This includes the operations of many of the world's leading state-sponsored groups, many foreign intelligence agencies across the world, and the top cyber-capable nations worldwide, which are able to execute such operations more than 100 times per year.

SL5 A system that could plausibly be claimed to thwart most top-priority operations by the top cyber-capable institutions (OC5).

This includes the handful of operations prioritized by the world's most capable nation-states.

Another consideration for structuring our recommendations into security levels is the ongoing development of responsible scaling policies, preparedness frameworks, and other similar voluntary policies for assessing and mitigating risks from advanced AI models. These voluntary policies describe mitigations that AI companies commit to implement to reduce the risk of harms from their models. A key component of these mitigations is implementing security measures that are proportionate to the risk presented by a model. These levels (sometimes termed *AI safety levels* or *risk levels*) are often defined by which threat actors the model should be secured against. Despite security measures being a critical component, these other frameworks have not yet defined concrete plans for security measures. The security levels in this report align with existing voluntary policies and provide a rough blueprint for AI companies to consider following as they further develop their policies.

That being said, we do not aim to be fully prescriptive. Our security level benchmarks represent neither a complete standard nor a compliance regime—they are provided for informational purposes only and should inform security teams' decisions rather than supersede them.

Security Level Benchmarks

Accurately assessing which real-world systems conform to the different security requirements is a challenging and even controversial task because of the specific circumstances of different systems, significant disagreements in the field, and the field's ever-changing nature. Security is a complex and nuanced endeavor, and recommendations must be adjusted to fit the circumstances and constraints of the specific system. However, providing AI organizations with benchmarks informed by the views of a diverse set of experts as a starting point to adapt as necessary can empower AI organizations to better protect themselves.

With that goal in mind, we describe a benchmark system for each security level, listing concrete measures and policies that are estimated to make up the minimum requirements of a system that conforms to the goals of that security level.

We considered five primary environments in which AI models need to be protected:

- model training
- research environment (focusing primarily on research done with trained large-scale models—for example, fine-tuning—and not research done on smaller models)
- internal deployment (for the organization's own use)
- public API deployment (including inference, outputting token likelihoods [known as "logprobs"], and customer fine-tuning)
- on-premises deployment (where a separate organization runs the model independently).

The security level benchmark systems were compiled and developed through iterative consultation with experts. The benchmarks will likely require corrections and adjustments over time; they aim to provide basic calibration, offer organizations concrete suggestions for next steps, and catalyze discourse around this topic. We welcome discussions on what protections should be included in each level.

Below, we provide a high-level description of the benchmarks for each of the five security levels. The full details of the security measures in the security levels can be found in Appendix B.

Security Level 1 (SL1)

Table 6.1 provides a brief overview of the SL1 benchmark.

Definition

A system that can likely thwart amateur attempts (OC1). This includes the operations of many hobbyist hackers, as well as more experienced hackers who implement completely untargeted "spray and pray" attacks.

TABLE 6.1
Overview of the Security Level 1 Benchmark

Category	Subcategory	Security Measure Labels
Weight Security	Weight Storage	• Sensitive data remain internal. • Weight encryption (best effort)
	Physical Security	• Data centers of cloud providers
	Access Control	• Access control for sensitive assets • Access log or audit trail
Security of Network and Other (Nonweight) Sensitive Assets	Software	• Moderately frequent software update management and compliance monitoring
	Access, Permissions, and Credentials	• Least privilege principle • Restrictions on device and account sharing • Password best practices • Multifactor authentication • Single Sign-On (SSO) • Backup and recovery tools • Commercial identity and access management (IAM) tools • Zero Trust architecture (adherence to at least the standards in the "Traditional" level of the Cybersecurity and Infrastructure Security Agency's [CISA's] Zero Trust Maturity Model)[a]
	Hardware	• Modern device architectures that establish root of trust and block malicious code execution • CPU anti-exploitation features
	Supply Chain	• The reputability of software is reviewed before incorporation.
	Security Tooling	• Modern authentication infrastructure • Commercial network security solutions • Commercial endpoint security solutions • Reliance on standard security infrastructure (depending on circumstances)
	Configuration Management	• Enforce screen locks for inactivity
Personnel Security	Awareness and Training	• Basic onboarding information security training for employees
Security Assurance and Testing	Risk and Security Assessments	• Internal reviews
Security Team Capacity	–	• Basic incident response capabilities
Maintenance	–	• Information security news monitoring and implementation

[a] CISA, *Zero Trust Maturity Model*, version 2.0, April 2023a.

Bottom Lines

- Organizations benefit from a variety of protections simply by using modern platforms and cloud providers.
- At this level of investment, it is better for organizations to rely on existing security products and best practices rather than trying to develop their own proprietary solutions, as attempting the latter is more likely to introduce unintended vulnerabilities than fix known ones.
- Organizations can readily enhance their security at this level by adopting stricter security configurations and policies. However, SL1 is insufficient to achieve reliable security against all but the most trivial attackers.

Security Level 2 (SL2)

Table 6.2 provides a brief overview of the SL2 benchmark.

Definition

A system that can likely thwart most professional opportunistic efforts by attackers that execute moderate-effort or nontargeted attacks (OC2). This includes the operations of many professional individual hackers, as well as capable hacker groups when executing untargeted or lower-priority attacks.

TABLE 6.2
Overview of the Security Level 2 Benchmark

Category	Subcategory	Security Measure Labels
Implementation of Previous Security Levels	–	• The organization has implemented all controls from SL1.
Weight Security	Weight Storage	• Storage location (e.g., weights are stored exclusively on servers and not on local devices) • Encryption (e.g., all keys are secured in a key management system)
	Security During Transport and Use	• Encryption in transit (i.e., not transporting weights over public or unencrypted channels)
	Physical Security	• Data centers are guarded, and only people with authorization are allowed inside. • Visitor access is restricted and logged.
	Access Control	• Restrictions on sensitive interactions (e.g., require multifactor authentication using FIDO authentication/hardware security keys)
	Monitoring	• Logging of all sensitive interactions • Regulation and monitoring of weight copies across the organization network
AI Model Resilience	Model Robustness	• Input reconstruction (e.g., during inference, a privately known prefix is added ahead of the user prompt) • Adversarial training
Security of Network and Other (Nonweight) Sensitive Assets	Software	• Frequent software update management and compliance monitoring
	Access, Permissions, and Credentials	• Strong password enforcement • The work network is separate from the guest network. • Guest accounts disabled whenever possible • Strong access management tools • Zero Trust architecture (the organization adheres to at least the standards in the "Initial" level of CISA's Zero Trust Maturity Model)[a]
	Hardware	• Lost or stolen devices reported • All network devices are visible and trackable.
	Supply Chain	• Review of vendor and supplier security
	Security Tooling	• Disk encryption • Network communications are encrypted by default. • Email security tools • Use of integrated security approaches
	Configuration Management	• Incorporate fundamental infrastructure and policies for Security-by-Design and Security-by-Default. • Configuration management monitoring
	Physical Security	• Office security • Careful disposal of printed materials

Table 6.2—Continued

Category	Subcategory	Security Measure Labels
Personnel Security	Awareness and Training	• Periodic mandatory information security training for all employees • Employee training on configuration errors and their security implications
	Filtering and Monitoring	• Installation of monitoring software for secure network access • Active drills to identify and educate noncompliant employees
Security Assurance and Testing	Red-Teaming and Penetration Testing	• Mandatory external reviews
	Community Involvement and Reporting	• Bug-bounty and vulnerability-discovery programs
	Software Development Process	• Secure software development standards (compliance with NIST's Secure Software Development Framework)
Incident Response	–	• Protocols and funding for rapid incident response • Incident reporting
Security Team Capacity	–	• Constant availability of qualified personnel
Maintenance	–	• Continuous vulnerability management and adaptation to information security developments
Other Organization Policies	–	• Promotion of a security mindset by organization management • Stringent remote work policies

[a] CISA, 2023a.
NOTE: FIDO = Fast Identity Online.

Bottom Lines

- This level consists primarily of implementing the fundamentals and latest industry best practices, comprehensively across the board. The most important concern at this level is ensuring that there are no "blind spots" left unaddressed. Prioritizing the most common attack vectors is key: for example, ensuring that email security, password policies, and multifactor authentication are all enforced correctly.
- Implementing security fundamentals and policies that improve security throughout the network, codebase, or organization is particularly critical. This includes robust and well-supported corporate infrastructure, secure software development framework, and zero-trust architecture. Multiple experts highlighted these fundamentals as particularly critical not only in the context of SL2 but also for robust security more generally.
- Red-teaming and other similar endeavors aimed at testing one's security system are an essential practice that first appear in SL2. The practices in this level are relatively basic but become increasingly rigorous at higher security levels.
- Centralizing all copies of the weights to a limited number of access-controlled and monitored systems is also a critical component, because it is a prerequisite for many other forms of security and control.

Note that the threat model in this level does not include insider threats, which are one of the biggest risks for a system at this level.

Security Level 3 (SL3)

Table 6.3 provides a brief overview of the SL3 benchmark.

Definition

A system that can likely thwart cybercrime syndicates or insider threats (OC3). This includes the operations of many world-renowned criminal hacker groups, well-resourced terrorist organizations, disgruntled employees, and industrial espionage organizations.

TABLE 6.3
Overview of the Security Level 3 Benchmark

Category	Subcategory	Security Measure Labels
Implementation of Previous Security Levels	–	• The organization has implemented all controls from SL1 and SL2.
Weight Security	Weight Storage	• Centralized and restricted management of weight storage • Secure cloud network (if applicable) • Dedicated devices for weights and weight security data
	Physical Security	• Data centers are guarded or locked at all times. • Premises are swept for intruders frequently. • Premises are meticulously swept for unauthorized devices routinely.
	Permitted Interfaces	• Authorized users who interact with the weights do so only through a software interface that reduces risk of the weights being illegitimately copied. • Any code accessing the weights minimizes attack surface, provides only simple forms of access, and uses the minimal amount of (highly trusted and well-established) external code necessary. • Avoiding model interactions that bypass monitoring or constraints
	Access Control	• Protocols and policies for sensitive interactions (e.g., access to the various permitted interfaces to the weights is stringently controlled, multiparty authorization, security reviews, etc.)
	Monitoring	• Ongoing manual monitoring of sensitive interactions • Ongoing automated anomaly detection • Automated and manual monitoring/blocking of potentially malicious queries • Frequent compromise assessment • Frequent integrity checks via comparison against a baseline system configuration ("gold image")
	Standard Compliance	• Implementation of measures described by NIST SP 800-171 or equivalent • Future implementation of measures described by CMMC 2.0 Level 3[a]
AI Model Resilience	Model Robustness	• Adversarial input detection
	Oracle Protection	• Limitations on the number of inferences using the same credentials
Security of Network and Other (Nonweight) Sensitive Assets	Software	• Very frequent software update management and compliance monitoring
	Access, Permissions, and Credentials	• 802.1x authentication • Zero Trust architecture (adherence to at least the standards in the "Advanced" level of CISA's Zero Trust Maturity Model)[b]
	Hardware	• Security-minded hardware sourcing
	Supply Chain	• Software inventory management • Supply chain security is commensurate with the organization's security
	Security Tooling	• Enforcement of security policies through code rather than manual compliance • Security policy enforcement for network access across devices

Table 6.3—Continued

Category	Subcategory	Security Measure Labels
Personnel Security	Awareness and Training	• Employee awareness of weight interaction monitoring • Security training for employees (not necessarily only those with access) • Security risk reporting program
	Filtering and Monitoring	• Insider threat program
Security Assurance and Testing	Red-Teaming and Penetration Testing	• Ongoing penetration testing • Penetration testing of physical access and facility security • Advanced red-teaming: - Elite external team - Substantial funding - Access to design and code - Testing insider threats - Expanded access - Attention to the weights and authentication
	Risk and Security Assessments	• Keeping a risk register
Threat Detection and Response	–	• Placement of effective honeypots
Security Team Capacity	–	• General increased capacity (compared with SL2) • Concrete experience with APTs • Leveraging diverse security experience from leading organizations
Other Organization Policies	–	• Two independent security layers

[a] Cybersecurity Maturity Model Certification (CMMC) 2.0 Level 3 ("Expert") requirements are defined by Chief Information Officer, U.S. Department of Defense, "CMCC Model," webpage, undated.

[b] CISA, 2023a.

Bottom Lines

- Aggressively reducing the attack surface is the key theme in this security level. To quote Mark Dowd, "[With a persistent attacker], the attack surface is the vulnerability. Finding a bug there is just a detail."[5]
- No less important: At SL3, one must assume that attackers have unexpected access or capabilities—because they are an insider, have zero-days, or spent more time researching the system than its developers. These capabilities drive many of the new requirements, including the critical requirements for defense-in-depth.
- A key goal of SL3 is reducing the risks from insider threats (e.g., company employees), which simultaneously reduces the risk from attackers who can masquerade as insiders or gain illegitimate access to employees' digital devices. One critical component of mitigating these risks is reducing the number of people with authorized access to the weights, detailed in the "Access Control" section of the SL3 description in Appendix B. Another critical component is hardening the interfaces to weight access against weight exfiltration, for which more detailed implementation recommendations are provided in Appendix B, in the "Permitted Interfaces" section of the description of SL3. Finally, another foundational requirement is the implementation of defense-in-depth, described in the "Other Organization Policies" section of the description of SL3.
- The full supply chain, be it software, hardware, or the air conditioners, is monitored and secured.
- The set of actors within OC3, the group that SL3 aims to protect against, is more diverse than in other categories—most notably in the inclusion of both insider threats and external cyber organizations. We group the OC3 actors together because the levels of investment required to robustly defend against them

[5] Mark Dowd, "How Do You Actually Find Bugs?" keynote video, April 21, 2022.

are comparable, despite the specific measures required being partially but not fully overlapping. In the SL3 benchmark, we include the union of measures required to protect against actors in this group, thus providing a recommendation for achieving the security assurance needed despite its diversity.

Security Level 4 (SL4)

Table 6.4 provides a brief overview of the SL4 benchmark.

Definition

A system that can likely thwart most standard operations by leading cyber-capable institutions (OC4). This includes the operations of many of the world's leading state-sponsored groups, many intelligence agencies across the world, and the top cyber-capable nations worldwide, which are able to execute such operations more than 100 times a year.

TABLE 6.4
Overview of the Security Level 4 Benchmark

Category	Subcategory	Security Measure Labels
Implementation of Previous Security Levels	–	• The organization has implemented all controls from SL1–SL3.
Weight Security	Weight Storage	• Isolation of weight storage • Weight storage setup is protected against eavesdropping and the simplest of TEMPEST attacks. • Hardware-enforced limits on output rate • Reduced communication capabilities
	Security During Transport and Use	• Confidential computing (when available)
	Physical Security	• Increased guarding (compared with SL3) via manned and digital systems • Meticulous logging of all access • Prohibiting devices near the setup
	Permitted Interfaces	• Specialized hardware for all external interfaces
	Monitoring	• Enforcement of time-buffered review (software limitation) • Protection of the monitoring logs at the hardware level • Comprehensive anomaly detection and alert system over the monitoring logs
AI Model Resilience	Model Robustness	• Adversarial output detection
	Oracle Protection	• Output reconstruction
Security of Network and Other (Nonweight) Sensitive Assets	Software	• Limiting the attack surface (e.g., the limited interaction interfaces of a Chromebook)
	Access, Permissions, and Credentials	• Enforcement of strong random passwords and keys for enhanced security • Zero Trust architecture (adherence to at least the standards in the "Optimal" level of CISA's Zero Trust Maturity Model)[a]
	Hardware	• All hardware used on devices must undergo source-code auditing and be validated as secure. • Secure hardware required for access • Ongoing compromise assessment on all devices with access
	Supply Chain	• Strict application allowlisting (especially for sandboxes) • SLSA Level 3 specification for all software used
	Security Tooling	• Significant investment in advanced security systems
	Physical Security	• Banning of unauthorized devices

Table 6.4—Continued

Personnel Security	Filtering and Monitoring	• Preventing third-party access and reporting suspected illegitimate incidents • Advanced insider threat program • Occasional employee integrity testing
Security Assurance and Testing	Red-Teaming and Penetration Testing	• Ongoing research and red-teaming to identify potential attack methods on the weight interface(s) • Ensuring physical security through red-teaming • Experience dealing with intelligence agencies
	Risk and Security Assessments	• Automated weight exfiltration attempts • Manual weight exfiltration attempts • Compliance with the FedRAMP High standards for security[b]
Security Team Capacity	–	• General increased capacity (compared with SL3) • Greater concrete experience with APTs (compared with SL3) • Zero-day vulnerability discovery capabilities • The security team is empowered to not compromise security over other stakeholders.
Other Organization Policies	–	• Designating sensitive details of the weight security system • Vetting of investors and other positions of influence • Prioritizing leak prevention over other organizational goals • Four independent security layers

[a] CISA, 2023a.

[b] See FedRAMP, "Understanding Baselines and Impact Levels in FedRAMP," blog post, November 16, 2017.

Bottom Lines

- Attack surfaces need to be reduced to the extent that the remaining security-critical surface can be comprehensively (and often manually) hardened, reviewed, monitored, and penetration tested. Such reduction requires significantly more compromises on productivity, convenience, and efficiency than previous levels.
- For security-critical junctions, any software assurances or general-purpose hardware are no longer considered trusted. Critical security assumptions need to be implemented in hardware. This may require changes to how data centers are set up.
- One important security measure in SL4 is the implementation of confidential computing to protect the weights in use. In Appendix B, we provide additional details on how this technology should be used to ensure that the weights are secure.
- Because state actors have extensive capabilities unavailable to other actors, the security team must have specific experience and expertise dealing with such actors; this requirement heavily influences how security assessment, red-teaming, and other activities are performed.
- Access to large numbers of zero-days and other capabilities that may be years ahead of public knowledge means that many security redundancies are needed (four independent security layers).

Security Level 5 (SL5)

Table 6.5 provides a brief overview of the SL5 benchmark. Please note that we believe it is not currently feasible to defend an internet-connected system against a determined and capable state actor using off-the-shelf solutions and standard industry principles. Hence, the multiple requirements listed in Table 6.5 need to be developed, though we avoided solutions that require novel research or for which feasibility is unknown.

Definition

A system that could plausibly be claimed to thwart most top-priority operations by the top cyber-capable institutions (OC5). This includes the handful of operations prioritized by the world's most capable nation-states.

TABLE 6.5
Overview of the Security Level 5 Benchmark

Category	Subcategory	Security Measure Labels
Implementation of Previous Security Levels	–	• The organization has implemented all controls from SL1–SL4.
Weight Security	Weight Storage	• Extreme isolation of weight storage (completely isolated network) • Advanced preventive measures for side-channel attacks (e.g., noise injection, time delays, and other tools) • Formal hardware verification of key components
	Physical Security	• Increased significant guarding (compared with SL4) • Supervised access for everyone • Routine rigorous device inspections • Disabling of most communication at the hardware level
	Permitted Interfaces	• Strict limitation of external connections to the completely isolated network
	Access Control	• Irrecoverable key policy (barring alternative access or key retrieval systems)
	Standard Compliance	• Protection equivalent to that required for Top Secret (TS)/Sensitive Compartmented Information (SCI)
AI Model Resilience	Oracle Protection	• Constant inference time
Security of Network and Other (Nonweight) Sensitive Assets	Supply Chain	• Strong limitations on software providers (e.g., only developed internally or by an extremely reliable source) • Strong limitations on hardware providers (e.g., only developed internally or by an extremely reliable source)
Personnel Security	Personal Protection	• Proactive protection of executives and individuals handling sensitive materials
Security Assurance and Testing	Red-Teaming and Penetration Testing	• Proactive search for crucial vulnerabilities (e.g., zero-days)
Maintenance	–	• Security is strongly prioritized over availability (e.g., barring connecting external devices to completely isolated network to debug a critical production issue).
Other Organization Policies	–	• Eight independent security layers

Bottom Lines

- Except for production use, weights are stored in a completely network-isolated setup (disconnected from the external world), with extremely stringent policies on data transfer that would prevent even those with approved access from being able to take large amounts of data out of the room.
- More R&D is needed to enable organizations to support production models while meeting SL5 security requirements. We recommend the development of hardware security module (HSM)-like devices with an interface that is specialized for ML applications.
- Achieving SL5 is currently not possible. Realizing all SL5 measures will likely require assistance from the national security community.

Notable Areas of Disagreement and Consensus

As with other topics, we encountered a diversity of opinions related to how best to secure against threats. Below, we describe which aspects of securing against threats manifested more or less variation in opinion:

- Variation in opinion increased from lower security levels to higher ones. Specifically:
  - There was significant convergence on the benchmarks for SL1 and SL2.
  - There was variation in opinion about the details of the SL3 benchmarks: Different experts maintained that SL3 security should be slightly higher in general, slightly lower in general, or that a small number of measures should be added or removed.
  - There are deeper and more conceptual disagreements about what is needed to achieve the security implied by SL4 and SL5—with opinions ranging from the SL3 benchmark being sufficient to secure against all threat actors to claims that no system could ever present a significant hurdle to operations in the OC5 category.
  - Among security measures in SL4 and SL5, those that were expected to result in significant costs to productivity, collaboration, and culture were the most controversial. There was significant variance in estimates of how beneficial, necessary, or costly such measures would be.
- A particular point of disagreement was the number of people who should have authorization to access the weights. Some experts strongly asserted that the model weights cannot be secure if this number is not aggressively reduced (e.g., to the low tens); others claimed that such a reduction would not be necessary, feasible, or justified.
  - The trade-off between security and productivity that this number of people points to can be significantly mitigated by implementing more constrained and secure interfaces for model weight access. For concrete recommendations on how to implement these, see the "Permitted Interfaces" section of SL3 and the "Weight Storage" sections of SL4 and SL5 in Appendix B.
- There was also heated debate regarding both the effectiveness and costs of rate-limiting outputs of systems with access to the weights: Some experts argued this is an incredibly useful prevention tool that can be implemented without significant limitations to legitimate use, whereas others claimed that doing so is both ineffective and cannot be done without severe harm to legitimate use.
- Similar to expert views on AI-specific attack vectors, there was a general recognition of significant uncertainty regarding the effectiveness of many AI-specific mitigations because the field is still nascent and evolving. As a result, we expect recommendations in this space to change rapidly over the coming years. We also refrained from including mitigations with little evidence of effectiveness, such as model pruning and network distillation—though they may turn out to be useful.

- There was more agreement on which measures should be included in each benchmark than which measures are doing most of the "heavy lifting." Across the security measures, experts varied significantly in which measures they assessed to be most significant, and, as with other aspects, diversity of opinion increased at higher security levels.
- Two points of overwhelming agreement were raised independently by many experts and agreed on by others:
  - The importance of modern and well-implemented security foundations and infrastructure, which allow security teams to control and monitor the relevant environments, implement other security measures more effectively, and respond to incidents.
  - Confidential computing as a strategic next step toward improving model weight security. Correct implementation of confidential computing for this purpose would be a major security improvement and is feasible in the near term—though still nascent and not production-ready as of this writing in mid-2024. For concrete recommendations on how to implement confidential computing to ensure the security of model weights, see the "Security During Transport and Use" section of SL4 in Appendix B.

Concluding Remarks for the Security Levels

The security levels and their benchmarks offer a rough tool for calibrating the relationship between the implementation of security measures and expected security outcomes.

Although the optimal security setup for a specific organization may be different from the benchmark security setup for a given level, each benchmark can assist in identifying whether an organization's security posture consistently falls short of experts' expectations for that specific security level (as opposed to achieving similar goals in a slightly different way), and therefore may not secure against the relevant operational capacity. Furthermore, conceptually distinct attack vectors (such as those belonging to separate attack categories) can often act as independent paths toward reaching the weights. For example, even if a system's security is otherwise comprehensive but additional (less secure) copies of the weights exist, the weights can still be stolen. The same applies to a vulnerability in a model distillation or extraction attack, and others.

Additional investment in securing against one category—for example, Running Unauthorized Code—does not compensate for gaps in securing against another, such as Unauthorized Physical Access to Systems or Human Intelligence. Therefore, security systems need to address the range of attack vectors discussed in this report. We generally recommend that organizations implement mature and effective versions of the controls in one level before investing the resources and effort required for the components of the next security level: A review of the security measures included quickly reveals that the cost and effort required rise significantly as the security levels increase. However, if an organization needs to achieve multiple security levels in a short time span (e.g., a few years), it may need to identify measures in more advanced levels that cannot be implemented within short timelines and begin working on these in parallel to implementing measures in lower security levels.

It was generally understood by many of the experts we spoke with that many organizations, not just those developing frontier models, have already implemented many of the security mechanisms outlined in this report, especially at the lower security levels. We believe it is important for organizations developing models with unprecedented capabilities to have a clear plan for securing models that are more capable than the current state of the art. We hope this report can help organizations identify which security goals they are already addressing and focus on those they have yet to address.

CHAPTER 7

Conclusion

Progress in AI capabilities presents both remarkable opportunities and significant challenges. Once a malicious actor possesses a model's weights, the barrier to misusing its associated capabilities without restrictions is low. Given the rapid rate of change for such capabilities, there is a need for robust, forward-thinking strategies to ensure the security of AI systems. Our discussion has underscored the complexity of securing such systems, specifically in the context of securing the weights of frontier AI systems from theft and, especially, against advanced threats.

This report offers four major contributions intended to help organizations meet the challenges of creating or improving security strategies:

1. defining operational capacity categories of attackers
2. identifying diverse attack vectors
3. estimating the feasibility of each attack vector being executed by the different capacity categories
4. defining security levels aimed at protecting from increasingly capable categories of malicious actors.

These contributions enable organizations to derive meaningful observations and decisions, such as identifying eight attack vectors that may not be on the radar of many companies, because security measures for such vectors are not needed when defending against less capable adversaries but are critical when defending against highly capable ones (OC4 and OC5). Furthermore, the categorization of operational capacities, alongside the feasibility of the vectors and the benchmark systems incorporated in the security levels, allows organizations to concretely assess whether they roughly meet the threshold of a security level and identify priorities and next steps for improving their security posture.

As is readily apparent from the benchmark systems, achieving higher security levels presents challenges, possibly necessitating compromises in operational efficiency. For example, the SL5 benchmark requires aggressively limiting how AI weights can be accessed, rethinking how data centers are built, investing heavily in security assurances and redundancies, and more. Some of these efforts may require years to implement. The security levels do not, in and of themselves, imply what security outcomes are *required*. They simply assist in calibrating between security measures implemented and security outcomes that are likely to be achieved.

Our discussion draws from a diverse range of sources, including academic research, industry reports, governmental publications, and interviews with 31 leading global experts. However, although the conclusions stem from extensive consultation, we see this report as only a first step toward a more rich and robust discourse on the security of AI systems. The complexity of the domain and its ever-evolving nature demand an ongoing conversation.

Additionally, we believe there is much more work to be done to promote the emerging field of AI security, including analysis of protecting the confidentiality, integrity, and availability of the various critical components of AI systems beyond their weights; development of a robust R&D agenda to expand the AI security toolkit; and better defining the roles of different actors in securing AI systems, among others. We sincerely hope this baseline effort will help to advance these important needs.

Detailed List of Attack Vectors

In this appendix, we provide detailed descriptions of the attack vectors, by attack category, as listed in Table 5.1 in Chapter 5, alongside examples of the attack vectors successfully used to penetrate real-world sensitive systems and additional context. The attack vectors are divided into conceptual categories primarily for readability.

Running Unauthorized Code

A ubiquitous attack vector in today's information security landscape is the exploitation of software vulnerabilities or accessible interfaces by malicious actors to run unauthorized code. Essentially, this means making a system or device execute commands dictated by an attacker rather than by its legitimate owner. This category includes methods that might not introduce new code into the system, such as living-off-the-land techniques (e.g., return-oriented programming). Running malicious code overlaps considerably with other categories: It is a key tool in such attacks as stealing credentials, supply chain attacks, and undermining access control systems, and many other types of attacks can be used to achieve remote code execution. For a more detailed categorization of vulnerabilities, refer to NIST's National Vulnerability Database categories.[1]

There are two conceptually distinct scenarios in this context:

1. **Target-specific vulnerabilities:** An attacker first identifies a concrete interface or piece of software relevant to a specific target device or network, then acquires a vulnerability for that target. Examples:
   - The Triangulation attack on Kaspersky employees was carried out by infecting employee iPhones with spyware, showing that these vulnerabilities are present even in companies that strongly prioritize security.[2]
   - Operation Aurora, a cyberattack that compromised trade secrets of leading U.S. companies, including Google, was accomplished via a targeted phishing campaign.[3]
   - Penetration testing efforts yield such vulnerabilities on a regular basis.
2. **Ubiquitous vulnerabilities:** An attacker acquires a vulnerability for a widely used piece of software or protocol that is likely to be broadly relevant to the target system, regardless of the system's specific details. Such vulnerabilities are much more valuable and sought-after and are accordingly harder to find. However, because of their broad applicability, larger-capacity groups (such as state actors and other large organizations) will invest substantial resources to find such vulnerabilities and use them across multiple operations. A sufficiently skilled actor would be capable of discovering these weak-

1 NIST, "NVD CWE Slice," National Vulnerability Database, August 3, 2023b.

2 Eugene Kaspersky, "A Matter of Triangulation," *Kaspersky Daily*, June 1, 2023.

3 Council on Foreign Relations, "Operation Aurora," webpage, undated.

nesses and highly incentivized to look for them, and the marginal costs of exploiting them might be very low since the fixed costs of finding them have already been paid. However, using them does pose some risk of compromising a vulnerability that could be useful in multiple operations. Examples:

- The WannaCry ransomware attack had global repercussions, infecting 200,000 computers across 150 countries, including a reported 40 percent of health care organizations and 60 percent of manufacturing organizations.[4]
- The NotPetya wiper attack had wide-ranging effects on many organizations, with estimated costs exceeding $10 billion in aggregate.[5]

To provide a sense of scale and frequency of generic vulnerabilities, almost 30,000 vulnerabilities with a Common Vulnerability Scoring System (CVSS) score of 9 or higher (defined as critical vulnerabilities, often enabling an arbitrary unauthorized attacker to gain control of a system) were reported over the past decade.[6] As of January 2024, AttackerKB,[7] a crowdsourced resource on vulnerabilities, indicates at least 186 vulnerabilities discovered in 2022 alone were exploited in the wild; of those, 47 had high or very high value to attackers.[8]

To understand how this reflects on ubiquitously used systems (as opposed to niche products), one can explore the vulnerability history of OpenSSL—a leading implementation of Transport Layer Security (TLS), the protocol that almost all internet users and servers use to secure their internet communications.[9] In the nine years since the severity of vulnerabilities has been classified, 21 high-severity vulnerabilities have been reported. As the examples below illustrate, many actors consistently identify *zero-days* (vulnerabilities that have not yet been identified or mitigated by the vendor or the broad cybersecurity community; that is, there have been at most "zero days" since the vendor discovered or mitigated the vulnerability) years before they are reported. Thus, we should expect capable actors to have access to multiple unreported vulnerabilities at a time.

Attackers can acquire such vulnerabilities in several ways, depending on available resources. The categories described below are roughly ordered from easiest and least valuable to hardest to acquire and most valuable.

Exploiting Vulnerabilities for Which a Patch Exists (Attacking Non-Updated Software)

These are known vulnerabilities that have been reported and have working patches (updates which fix the vulnerability), but some systems remain unpatched because of lax update policies or because of the use of products for which patches are no longer being provided by the vendor. Examples:

---

[4] Cloudflare, "What Was the WannaCry Ransomware Attack?" webpage, undated-b; Armis Security, "Two Years In and WannaCry Is Still Unmanageable," May 29, 2019.

[5] Andy Greenberg, "The Untold Story of NotPetya, the Most Devastating Cyberattack in History," *Wired*, August 22, 2018b.

[6] NIST, 2023c; MITRE, "CVSS Scores Between 2013-07-30 and 2023-07-30," CVEdetails.com, undated-c. These results were produced by searching the common vulnerability and exposure (CVE) details database for CVSS scores between July 30, 2013, and July 30, 2023.

[7] AttackerKB, homepage, undated-c.

[8] AttackerKB, CVE year 2022 search results, webpage, undated-b; AttackerKB, CVE year 2022 search results sorted by value, webpage, undated-a. The result of 186 vulnerabilities was produced by searching the AttackerKB database using the search tag "exploited in the wild" and limiting the year to 2022. The result of 47 high- or very high-value items was produced by limiting the list of 186 to an attacker value of 4.

[9] OpenSSL, "Vulnerabilities," webpage, undated.

- The SamSam ransomware used common vulnerabilities and exposures (CVEs) identified in 2010–2012 to target non-updated systems in the 2015–2018 time frame.[10] It had reportedly earned its creators $5.9 million by 2018.[11]
- The LastPass breach started with a code execution vulnerability that had been patched more than two years earlier (in a media software package on an employee's personal computer, further underscoring the large attack surface for finding such vulnerabilities).[12]
- Heartbleed was a bug in OpenSSL disclosed in 2014. Although it was reportedly exploited for months before its public disclosure,[13] and thus at early stages was used as a zero-day (see the next section), it is most famous for the immense impact it had even after disclosure, because many systems remained exploitable long after a patch was available.
- Penetration testing toolkits such as Metasploit and Cobalt Strike are consistently used by cyber attackers, allowing even actors with extremely limited expertise to systematically probe systems for a wide variety of known vulnerabilities.[14]

For an overview of how common system vulnerabilities are, see Qualys' SSL Pulse data.[15] At the time of this writing, approximately 40 percent of websites have known vulnerabilities or security issues detected by automated scans.

Exploiting Reported but Not (Fully) Patched Vulnerabilities

These are known vulnerabilities for which the vendor has either not provided a patch or provided a patch that only partially addressed the issue, primarily because of the challenge of patching the vulnerabilities comprehensively. During the window in which vulnerabilities remain unpatched, malicious actors can readily exploit them, because their existence is already known. Keeping one's software up to date with security patches is not sufficient protection against such vulnerabilities. Examples:

- Many VPN-SSL flaws, such as the ones identified in FortiGate's FortiNet products,[16] are known to be exploited quickly in the days after a patch is released to gain access into private networks.
- Spectre and Meltdown exploit the side effects of speculative execution in modern CPUs. Although it is uncertain how extensively these vulnerabilities have been exploited in the wild (e.g., the first weaponized exploit was only discovered in 2021), similar vulnerabilities have been found repeatedly, with a recent new batch discovered in August 2023.[17]

[10] "Shutting Out SamSam Ransomware," *Sophos News*, May 2, 2018.

[11] Sophos, *SamSam: The (Almost) Six Million Dollar Ransomware*, 2018.

[12] Michael Kan, "LastPass Employee Could've Prevented Hack with a Software Update," *PC*, March 3, 2023.

[13] "The Heartbleed Bug," webpage, June 3, 2020.

[14] Metasploit, homepage, undated; Fortra, "Software for Adversary Simulations and Red Team Operations," webpage, undated; Catalin Cimpanu, "Cobalt Strike and Metasploit Accounted for a Quarter of All Malware C&C Servers in 2020," *ZDNet*, January 7, 2021a.

[15] Qualys, "SSL Pulse," webpage, February 2, 2024.

[16] Lawrence Abrams, "Fortinet Fixes Critical RCE Flaw in Fortigate SSL-VPN Devices, Patch Now," *Bleeping Computer*, June 11, 2023.

[17] Graz University of Technology, "Meltdown and Spectre," webpage, undated; Catalin Cimpanu, "First Fully Weaponized Spectre Exploit Discovered Online," *The Record*, March 10, 2021b; "Collide+Power, Downfall, and Inception: New Side-Channel Attacks Affecting Modern CPUs," *Hacker News*, August 9, 2023.

- BREACH is an example of an attack that exploits an interaction between two foundational components of separate systems: compression in HTTP and encryption in TLS.[18] This broad applicability of BREACH, as well as the fact that mitigating it involved a significant hit to efficiency, led to BREACH not being comprehensively patched for years. Even after being patched, similar vulnerabilities due to application-layer compression periodically surface.
- Downgrade attacks are another common example of attacks that tend not to be comprehensively patched even when discovered.[19] Because large portions of the internet use older versions of software, many protocols must support older versions—even if they are known to have vulnerabilities. An attacker can intentionally force the usage of an older version and exploit a well-known vulnerability. Vendors often prioritize accessibility to out-of-date devices at the expense of the security of patched devices: The former often immediately affects revenues or user complaints, while the latter does not.
- Different variations on Bleichenbacher attacks were at different times zero-days, not fully patched, or fully patched. But they are a particularly good example of a type of vulnerability that, even once identified, is difficult to resolve completely given its broad relevance to RSA (i.e., the Rivest-Shamir-Adleman public-key encryption algorithm) and the possibility of minor variations and backward compatibility requirements.[20]
- Like the above, Vaudenay/POODLE type attacks have fallen under all the zero-day categories, but because of their broad relevance and minor variations, they evade comprehensive patching and periodically reappear.[21]
- One of the largest hacks in the history of the United Kingdom, reported in August 2023, used ProxyNotShell.[22] This vulnerability was originally exploited as a zero-day (see below), but we mention it here because it took Microsoft months to properly patch it despite its being exploited at large scales.
- PixieFail is a set of vulnerabilities in the TCP/IP stack of common open-source UEFI implementation discovered in January 2024,[23] allowing remote code execution with a local network access. Because many vendors use this implementation and each one needs to provide a fix independently, the issue was made public before all vendors (of unknown number) addressed it.

Finding and Exploiting Individual Zero-Days

Zero-days are vulnerabilities that are not yet identified by the vendor or the broader cybersecurity community. Therefore, a patch for them does not exist, and it is harder to detect their exploitation.

Individual zero-days can be bought in markets for prices between $10,000 and $2.5 million,[24] depending on the platform. One zero-day market platform claims to have received more than 15,000 submissions over its eight years of existence.[25]

[18] Angelo Prado, Neal Harris, and Yoel Gluck, "Breach Attack," webpage, undated.

[19] Bart Lenaert-Bergman, "What Are Downgrade Attacks?" *Crowdstrike*, March 14, 2023.

[20] Gage Boyle and Kenny Paterson, *20 Years of Bleichenbacher's Attack*, Royal Holloway University of London, ISG MSc Information Security thesis series, 2019.

[21] Nick Sullivan, "Padding Oracles and the Decline of CBC-Mode Cipher Suites," Cloudflare, blog post, February 12, 2016.

[22] Zack Whittaker, "Parsing the UK Electoral Register Cyberattack," *TechCrunch*, August 9, 2023; Carly Page, "Rackspace Blames Ransomware Attack for Ongoing Exchange Outage," *TechCrunch*, December 6, 2022.

[23] "PixieFail UEFI Flaws Expose Millions of Computers to RCE, DoS, and Data Theft," *Hacker News*, January 18, 2024.

[24] Zerodium, "Zerodium Exploit Acquisition Program," webpage, undated-b.

[25] Zerodium, homepage, undated-a.

Some attackers discover their own zero-days. Although the form of acquisition may not directly matter to someone securing a system, self-discovered zero-days generally can be stored or exploited for longer time frames than purchased ones without being publicly revealed. According to a 2017 RAND report,[26] the median time from initial detection of a zero-day by an exploit developer to detection by an outside party is five years (Figure 3.5). Examples:

- According to Google Project Zero's 0-days In-the-Wild collection,[27] since 2014, an average of 30 zero-days per year are detected as being exploited in the wild before they are officially discovered. Note that this is a lower bound on the number of actual zero-days exploited, because some are not detected.[28]
- NSO Group's iMessage exploit is an example of a nonstate actor developing a capability that for years allowed all their clients to run malware on any iPhone in the world without the individuals knowing they were targeted.[29] iPhones are generally considered more secure than most other platforms. NSO also chained together three zero-click exploits against iPhones in 2022,[30] and another of NSO's exploit chains was found in the wild in 2023.[31]
- Operation Triangulation used four zero-days to spy on Kaspersky employees' iPhones.[32]
- EternalBlue is an exploit of a server message block (SMB) vulnerability. Reports indicate that the exploit was discovered by the National Security Agency (NSA) years before it was known publicly and patched.[33]
- Stuxnet used at least four zero-days in its operations.[34]
- In 2017, the *Washington Post* reported that the Vault7 leak detailing Central Intelligence Agency (CIA) capabilities showed that, at the time of the leak, the CIA had zero-days that enabled sprawling access to exploits for mobile phones (iOS and Android), PC operating systems (Windows, MacOS, and Linux), routers (Linksys, D-Link, and others), web browsers (Chrome, Internet Explorer, Firefox, and Opera), and more.[35]
- Mandiant has reported that the number of zero-days exploited in the wild is at an all-time high, highlighting its use by China's Hafnium group.[36]

---

[26] Ablon and Bogart, 2017.

[27] Google Project Zero, 0-Days In-the-Wild, database, undated.

[28] For a review of 0-Days In-the-Wild in 2022, see Maddie Stone, "The Ups and Downs of 0-Days," Google Threat Analysis Group blog post, July 27, 2023.

[29] Ian Beer and Samuel Groß, "A Deep Dive into an NSO Zero-Click iMessage Exploit: Remote Code Execution," Google Project Zero blog post, December 15, 2021.

[30] Ravie Lakshmanan, "NSO Group Used 3 Zero-Click iPhone Exploits Against Human Rights Defenders," *Hacker News*, April 20, 2023b.

[31] "BLASTPASS: NSO Group iPhone Zero-Click, Zero-Day Exploit Captured in the Wild," Citizen Lab, Munk School of Global Affairs & Public Policy, University of Toronto, September 7, 2023.

[32] Boris Larin, "Operation Triangulation: The Last (Hardware) Mystery," Kaspersky SecureList, December 27, 2023.

[33] Ellen Nakashima and Craig Timberg, "NSA Officials Worried About the Day Its Potent Hacking Tool Would Get Loose. Then It Did," *Washington Post*, May 16, 2017.

[34] CISA, "Stuxnet Malware Mitigation (Update B)," Alert Code ICSA-10-238-01B, updated January 8, 2014.

[35] Miller and Nakashima, 2017.

[36] James Sadowski, "Zero Tolerance: More Zero-Days Exploited in 2021 Than Ever Before," Mandiant blog post, April 21, 2022.

Direct Access to Zero-Days at Scale

Capable state actors are also likely to have (official or surreptitious) access to information on not yet publicly disclosed zero-days for which a patch has not been officially released. For state actors with a large national information security community, gaining such access seems reasonably easy to legislate or put in place. For others, it would require surreptitious access to discussions with people who have strong information security backgrounds. Although this capability results in access to zero-days, as does the previous attack vector, this capability generates an enormous quantitative difference in the number of zero-days available to an actor, which in turn generates a qualitative difference in the threat model. Protecting against an actor that has access to 2 zero-days looks very different from protecting against an actor that has 50 zero-days. Many mitigations that can be effective in thwarting an attacker with access to a small number of zeros days, such as software-based monitoring systems and modest defense-in-depth practices (the use of multiple lines of defense), can be undermined or overcome by an actor with access to many zero-days. Examples:

- China has a law requiring disclosure of zero-days to the state before informing the responsible parties (e.g., the company that owns the vulnerable product) as part of a coordinated vulnerability disclosure.[37] They are also reported to use this information for active zero-day exploitation.[38] Hence, we should expect the Chinese government to have access to most zero-days reported from China before they are patched.
- Chinese APTs have gained access to NSA vulnerabilities multiple times years before they were publicly reported (or caught). One famous example is the EpMe vulnerability,[39] acquired and used at least four years before its discovery.
- The U.S. government has established a Vulnerability Equities Process to determine whether to disclose information about zero-days or withhold the information to exploit it.[40]

Compromising Existing Credentials

This category involves abusing credentials (such as encryption keys, passwords, cookies, etc.) that have been legitimately assigned to a user. It can be the full attack (gaining access to the model weights), or a stepping stone toward accessing tools or resources enabling a much more advanced attack. For example, compromising the credentials of an organization employee could allow the reset of rate-limit counters on a customer's API access, allowing model distillation attacks. Similarly, employees or third-party auditors could have privileged access to internal networks in which the next step of a cyberattack can be launched.

There are multiple ways an attacker could access or compromise such credentials, discussed below.

---

[37] Brad D. Williams, "China's New Data Security Law Will Provide It Early Notice of Exploitable Zero Days," *Breaking Defense*, September 1, 2021; Marleen Weulen Kranenbarg, Thomas J. Holt, and Jeroen van der Ham, "Don't Shoot the Messenger! A Criminological and Computer Science Perspective on Coordinated Vulnerability Disclosure," *Crime Science*, Vol. 7, November 19, 2018.

[38] Priscilla Moriuchi and Bill Ladd, *China's Ministry of State Security Likely Influences National Network Vulnerability Publications*, Recorded Future, 2017.

[39] Eyal Itkin and Itay Cohen, "The Story of Jian—How APT31 Stole and Used an Unknown Equation Group 0-Day," Check Point Research, February 22, 2021.

[40] Electronic Privacy Information Center, "Vulnerabilities Equities Process," webpage, undated.

Social Engineering

A relatively cheap and nonrisky toolset for influencing employees to perform actions that would allow an attack. Multifactor authentication is often considered an appropriate response to social engineering attacks, and most companies have multifactor authentication (see the description of SL1 in Appendix B). Yet slightly more advanced social engineering attacks can overcome the most common multifactor authentication schemes. This is why security keys, discussed in the description of SL2 in Appendix B, are often called phishing-resistant multifactor authentication.[41]

Automated/General Social Engineering Scheme

This scheme is extremely easy to execute, and even amateur attackers can do it at enormous scales. Yet it remains incredibly effective if there are no system-level defense mechanisms in place to prevent it. Examples:

- This is the most common component of attack in ISACA and Looking Glass's *State of Cybersecurity 2022* (and many other summary reports).[42] It is reportedly responsible for $6.9 billion in stolen funds in 2021 (the total may be higher).[43]
- Phishing-as-a-service that can overcome multifactor authentication is reportedly sold for a subscription of $400 per month; this specific service alone supports tens of thousands of phishing emails per month.[44]
- Proofpoint's *2022 Social Engineering Report* provides examples of coordinated phishing and social engineering, featuring the most common recent malware campaigns.[45]

More-Targeted Schemes (e.g., Spearphishing)

In this approach, the attacker adjusts the content to the individual (or organization) being targeted. Although slightly more tailored, spearphishing is still incredibly common and done at scale. More advanced targeting aimed at executives is often called *whaling*.[46] Although, historically, spearphishing could not be fully automated, it still requires only minimal investment per target. Furthermore, many offensive cyber approaches may be enhanced by modern AI tools, and we suspect spearphishing to be among the first, because existing LLMs are already automating this process.[47] Examples:

- In 2015, attackers used a coordinated spearphishing campaign to access the internal network of a power distribution company and gain control of part of the Ukrainian power grid by harvesting credentials from the account management system.[48]

---

[41] Bob Lord, "Phishing Resistant MFA Is Key to Peace of Mind," Cybersecurity and Infrastructure Security Agency, blog post, April 12, 2023.

[42] ISACA and Looking Glass, *State of Cybersecurity 2022: Global Update on Workforce Efforts, Resources and Cyberoperations*, 2022.

[43] Internet Crime Complaint Center, *Internet Crime Report 2021*, Federal Bureau of Investigation, 2021.

[44] "Cybercriminals Increasingly Using EvilProxy Phishing Kit to Target Executives," *Hacker News*, August 10, 2023.

[45] ProofPoint, *2022 Social Engineering Report*, 2022.

[46] Kaspersky, "What Is a Whaling Attack?" webpage, undated-b.

[47] Lily Hay Newman, "AI Wrote Better Phishing Emails Than Humans in a Recent Test," *Wired*, August 7, 2021.

[48] Kim Zetter, "Inside the Cunning, Unprecedented Hack of Ukraine's Power Grid," *Wired*, March 3, 2016.

- Uber's 2022 breach was accomplished using spearphishing alone.[49] The attacker also used social engineering to overcome multifactor authentication via an authentication app. This is a good example of using social engineering as a first step in a more advanced attack: The attacker used these credentials to access Uber's SentinelOne portal and run code on new machines and maintain persistence.
- The same attacker (allegedly) also breached Rockstar Games mere days later with a similar approach.[50]
- The Democratic National Committee was hacked multiple times through spearphishing.[51]
- The loosely organized group Lapsus$, at least some of whom are teenagers, has successfully gained access to sensitive information at Microsoft, Nvidia, Okta, Samsung and many other prominent companies using spearphishing, SIM swapping, and other tools in this category.[52]
- The Office of Personnel Management (OPM) breach, the largest breach of U.S. federal data in history, is thought to be the result of social engineering that provided access to legitimate credentials.[53]
- Using targeted SIM swapping methods, a criminal ring stole $400 million worth of cryptocurrencies from 50 U.S. individuals.[54]
- One in five organizations surveyed by Barracuda had accounts compromised due to spearphishing in 2021.[55]

Password Brute-Forcing and Cracking

Attackers masquerade as a user(s) with permissions thanks to password brute-forcing and cracking. This approach uses default passwords, common passwords from large database leaks, large numeration schemes, and common schemes that incorporate personal information (e.g., phone number, name). The efficiency of enumeration can vary wildly (depending on such circumstances as access to a hash of the password, web or local interface, and hash algorithm used), but it is often higher than one would naively expect due to some well-known algorithms for enumerating efficiently on passwords (e.g., using rainbow tables if no salt is used[56]).

This approach spans the full continuum of difficulty depending on the strength of passwords; however, if no policy is in place, it is highly likely that some employees will have extremely easy-to-guess passwords (6 percent of all passwords are from the 1,000 most common passwords, and 54 percent are from the top 10 million).[57] Password enumeration can be fully automated and done on a mass scale, with marginal cost per attack of dollars at most. Examples:

---

49 Corin Faife, "Uber's Hack Shows the Stubborn Power of Social Engineering," *The Verge*, September 16, 2022.

50 Siladitya Ray, "Social Engineering: How a Teen Hacker Allegedly Managed to Breach Both Uber and Rockstar Games," *Forbes*, September 20, 2022.

51 Calyptix Security, "DNC Hacks: How Spear Phishing Emails Were Used," blog post, December 30, 2016.

52 "Two LAPSUS$ Hackers Convicted in London Court for High-Profile Tech Firm Hacks," *Hacker News*, August 25, 2023; Kaspersky, "What Is SIM Swapping?" undated-c.

53 Brendan I. Koerner, "Inside the Cyberattack That Shocked the US Government," *Wired*, October 23, 2016.

54 Ashley Belanger, "SIM-Swapping Ring Stole $400M in Crypto from a US Company, Officials Allege," *Ars Technica*, January 30, 2024.

55 Barracuda MSP, *Spear Phishing: Top Threats and Trends*, Vol. 7, March 2022.

56 For an overview of these concepts, see Christophe Limpalair, "Hash Tables, Rainbow Table Attacks, and Salts," *Cybr*, July 11, 2022.

57 Ata Hakçıl (ignis-sec) and Oxflotus, "PWDB—New Generation of Password Mass-Analysis (Pwdb-Public)," GitHub, undated.

- According to Google Cloud, weak passwords are consistently exploited by state actors and account for nearly half of all security incidents on Google Cloud customers' accounts.[58]
- The 2014 Sony Hack used a worm that guessed passwords for SMB connections.[59]
- While simple password enumeration schemes can be prevented via multifactor authentication (see the description of SL1 in Chapter 6 and Appendix B), at least some implementations of multifactor authentication can be overcome by creative attackers, as was done by Lapsus$ and the SolarWinds hackers.[60]

Exploitation of Exposed Credentials

Authorized users may be careless with their credentials and leave them accessible to an attacker. Additionally, a system may be carelessly designed to allow surreptitious access to credentials even without any neglect by the user. For example, a suboptimally designed system may allow a person-in-the-middle attack whereby an attacker can arrange to access credentials as they are transported between the user and the legitimate server. Credentials can be compromised by the breach of other services or applications if credentials are shared across accounts. Examples:

- Passwords are often simply written down:[61] 57 percent of employees write passwords on sticky notes, 49 percent save passwords in plaintext documents, and 62 percent share passwords by SMS or email.
- Exposed private/symmetric keys or passwords on public GitHub repositories. Toyota and Microsoft both suffered breaches by exposing credentials in this way,[62] and GitHub has even implemented a free tool to scan repositories for exposed credentials.[63]
- A misconfigured server exposed credentials for Verkada internal systems in 2021, allowing access to customers' private data.[64]
- Dozens of plaintext Amazon Web Services (AWS) admin access keys belonging to companies and universities were discovered to have been inadvertently included in public packages of PyPi.[65]
- Tens of thousands of apps on the Google Play Store have hard-coded secrets that are easily accessed, such as API keys and unsecured databases.[66]
- 5.5 out of every 1,000 GitHub commits contain unprotected software secrets, such as seeds, API keys, and passwords.[67]

---

[58] Christopher Porter, ed., *Threat Horizons: April 2023 Threat Horizons Report*, Google Cloud Office of the CISO, April 2023.

[59] CISA, "Targeted Destructive Malware," Alert Code AA21-008A, updated January 3, 2020a.

[60] Dan Goodin, "Lapsus$ and SolarWinds Hackers Both Use the Same Old Trick to Bypass MFA," *Ars Technica*, March 29, 2022.

[61] Keeper Security, "Workplace Password Habits Leave Organizations Vulnerable to Cyber Attacks," webpage, 2021.

[62] Dwayne McDaniel, "Toyota Suffered a Data Breach by Accidentally Exposing a Secret Key Publicly on GitHub," *GitGuardian*, October 11, 2022; Joseph Cox, "Microsoft Employees Exposed Own Company's Internal Logins," *Vice*, August 16, 2022.

[63] Sergiu Gatlan, "GitHub Rolls Out Free Secret Scanning for All Public Repositories," *Bleeping Computer*, December 15, 2022.

[64] Tom Forbes, "I Scanned Every Package on PyPi and Found 57 Live AWS Keys," blog post, January 6, 2023.

[65] Forbes, 2023.

[66] "Thousands of Android Apps Leak Hard-Coded Secrets, Research Shows," *Cybernews*, September 1, 2022.

[67] Robert Lemos, "Inside Threat: Developers Leaked 10M Credentials, Passwords in 2022," *Dark Reading*, March 9, 2022.

Expanding Illegitimate Access (e.g., Escalating Privileges)

Existing credentials are often obtained by first gaining some illegitimate access to a system (e.g., using other tools described in this report) and then obtaining credentials that provide further access. A more indirect version of this approach is to purchase credentials from other disclosures. This could include passwords (or other credentials) that are used across accounts—for example, if employees use the same password for both their work and personal accounts, their personal account's password may be exposed in a major data leak, which can be used to access their work account. Examples:

- The Sands Hotel hack breached a not-well-protected test server and then extracted the credentials of a senior systems engineer via mimikatz to compromise the rest of the network.[68]
- The NotPetya global wiper attack also used mimikatz to extract user passwords out of RAM and access additional machines using the same credentials.[69]
- The LastPass breach used a vulnerability in software on an employee's personal computer to install a keylogger, and then used that employee's password to access sensitive LastPass material.[70] The LastPass breach itself is a case of using illegitimate access to access vast numbers of user passwords.
- The Triton/Trisys malware attacked multiple power plants over multiple years.[71] Credential theft malware was used to gain access to additional systems.[72]
- In the Target breach, attackers first infiltrated a third-party heating, ventilation, and air conditioning (HVAC) vendor, then used that vendor's access to Target's network to gain access to more sensitive areas storing consumer data.[73]
- Microsoft's code-signing keys, critical for ensuring that Windows computers do not run malicious code, have been repeatedly stolen or abused, then used to support further attacks.[74]
- Using an employee's reused password that was leaked in a LinkedIn breach, hackers gained access to Dropbox user credentials in 2012.[75]
- A compromised GoDaddy system password allowed hackers to gain original admin passwords and some SSL private keys of 1.2 million WordPress accounts.[76]

Especially vulnerable to the above types of attack are credentials or other forms of access for **administrative or compliance purposes**. In conversations with experts, we heard that even in systems with multiple

[68] Sean Gallagher, "Iranian Hackers Used Visual Basic Malware to Wipe Vegas Casino's Network," *Ars Technica*, December 11, 2014.

[69] Greenberg, 2018b.

[70] Dan Goodin, "LastPass Says Employee's Home Computer Was Hacked and Corporate Vault Taken," *Ars Technica*, February 7, 2023a.

[71] Steve Miller, Nathan Brubaker, Daniel Kapellmann Zafra, and Dan Caban, *TRITON Actor TTP Profile, Custom Attack Tools, Detections, and ATT&CK Mapping*, Mandiant, updated November 25, 2022.

[72] Steve Miller, Nathan Brubaker, Daniel Kapellmann Zafra, and Dan Caban, "Appendix B: Technical Analysis of Custom Attack Tools," in *TRITON Actor TTP Profile, Custom Attack Tools, Detections, and ATT&CK Mapping*, Mandiant, April 2019; via Internet Archive, stored on October 17, 2021.

[73] U.S. Senate, Committee on Commerce, Science, and Transportation, *A "Kill Chain": Analysis of the 2013 Target Data Breach*, majority staff report, 2014.

[74] Dan Goodin, "Microsoft Signing Keys Keep Getting Hijacked, to the Delight of Chinese Threat Actors," *Ars Technica*, August 25, 2023d.

[75] Samuel Gibbs, "Dropbox Hack Leads to Leaking of 68m User Passwords on the Internet," *The Guardian*, August 31, 2016.

[76] Zack Whittaker, "GoDaddy Says Data Breach Exposed over a Million User Accounts," *TechCrunch*, November 22, 2021.

layers of access control, encryption, and other defenses, certain individuals (e.g., in DevOps, the legal department, or law enforcement) may have unique "break glass" access to sensitive information (e.g., access to all relevant encryption keys) for administrative or compliance reasons or for an emergency. We are particularly concerned about this avenue because it bypasses many defense systems in place, it often provides access to incredibly sensitive elements that cannot easily be reached anywhere else, and some of these departments are not tech-savvy and more vulnerable to attack.

Undermining the Access Control System Itself

An attacker can use a variety of tools to undermine the mechanisms responsible for reliably providing (or withholding) access. We use the term *access control system* loosely to include any security component responsible for preventing unauthorized access to data, such as encryption, authentication, and the permissions system. Despite being more complicated and less common in discourse than some of the above attack vectors, undermining the access control system is sufficiently common to be ranked as the most frequent vulnerability category in the OWASP Top 10 (though, of course, the exact definitions may differ).[77]

Encryption/Authentication Vulnerabilities (in the Access Control System)

This vector entails breaking the existing encryption/authentication scheme underlying the security of the access control system. The most direct and straightforward (though not easy) way to overcome an access control system is to find a flaw in the system that enables one to do what the system was meant to prevent—for example, decrypt a secret, forge credentials, or get access to a resource despite not being an authorized user or having the relevant key. Such direct undermining of foundational pieces of a security system is usually done only by very advanced and capable actors.

A more common avenue for attack is to compromise a credentials authority, making it possible to generate illegitimate credentials, certificates, or other cryptographic assets. Attackers may achieve such a compromise in various ways (including running unauthorized code, compromising existing credentials, or other attacks listed in this report). However, once they have done so, they can undermine the broader authentication system in ways that may significantly exceed their previous access.

Acquiring this capability is very costly and quite rare. We expect that only a few capable state actors will be able to successfully attack foundational encryption schemes.

An important caveat is that the modern security system attack surface is very large—consisting of key generation, nonce generation, asymmetric authentication scheme, hash functions, symmetric encryption, key management, authentication token generation and validation, other (non-crypto) protocol elements, etc. So, while undermining a predefined component is extremely difficult, identifying the weakest link may be easier. However, once an actor has acquired this capability (depending on the capability, but at least in some circumstances), the cost of using it is vastly lower. Actors would still need to prioritize its use because each use increases the chance that it will be detected, but they could still easily use it 10,000 times per year. Examples:

- The WEP algorithm, which used to be responsible for all Wi-Fi encryption, had severe vulnerabilities,[78] the most critical of which was a vulnerability with the RC4 encryption algorithm itself.

[77] OWASP, "OWASP Top Ten," webpage, undated-b.

[78] Nikita Borisov, Ian Goldberg, and David Wagner, "Intercepting Mobile Communications: The Insecurity of 802.11," *MobiCom '01: Proceedings of the 7th Annual International Conference on Mobile Computing and Networking*, July 2001.

- In 2008, academics showed that MD5's vulnerability to collision attacks allowed the creation of fake certificate authorities,[79] undermining the security of TLS, the primary protocol responsible for secure communication over the internet.
- In 2012, it was discovered that a long-standing operation exploited MD5's vulnerabilities to forge a Microsoft-signed certificate using an MD5 collision to achieve a similar goal.[80]
- Differential cryptanalysis was a powerful technique that could undermine the security of most ciphers and hash functions at the time of its public discovery in the late 1980s. However, it later turned out that IBM had discovered it as early as 1974, at which point the NSA was already well aware of this technique and had used it to improve the security of the Data Encryption Standard, while not revealing that it had done so to the public.[81]
- In 2015, the Logjam vulnerability in Diffie-Helman sharing prime numbers was discovered, leading to the hypothesized vulnerability to state-level adversaries of a significant portion of TLS (internet browsing), VPN (secure network connection), and SSH (secure remote shell) connections globally.[82]
- As part of the SolarWinds breach, attackers were able to forge Microsoft cloud authentication tokens, compromising the cloud resources of the affected organizations.[83]

Intentional Backdoors in Algorithms, Protocols, or Products (in the Access Control System)

Backdoor is a term for flaws intentionally introduced into a security-related algorithm or protocol, rendering it insecure (either to anyone who identifies this flaw or only to those who have access to some backdoor key or secret). Intentional backdoors and supply chain attacks (described below) have significant overlap.

Intentional backdoors in popularly used infrastructure are extremely difficult to achieve for any actor without significant influence over standards or technological infrastructure (e.g., everyone except a small number of state actors and a somewhat larger number of large tech companies). For those actors, it is challenging but possible, as indicated by the NSA's mixed success rate (parts of which are now public due to the Snowden disclosures). Examples:

- Operation Rubicon was a covert operation in which the CIA and German intelligence secretly owned Crypto AG, manipulating the company's devices to weaken their encryption.[84]

---

[79] Alexander Sotirov, Marc Stevens, Jacob Appelbaum, Arjen Lenstra, David Molnar, Dag Arne Osvik, and Benne de Weger, "MD5 Considered Harmful Today: Creating a Rogue CA Certificate," Eindhoven University of Technology, Mathematics and Computer Science, December 30, 2008.

[80] Microsoft, "Flame Malware Collision Attack Explained," blog post, June 6, 2012.

[81] "The NSA's Work to Make Crypto Worse and Better," *Ars Technica*, September 6, 2013.

[82] David Adrian, Karthikeyan Bhargavan, Zakir Durumeric, Pierrick Gaudry, Matthew Green, J. Alex Halderman, Nadia Heninger, Drew Springall, Emmanuel Thomé, Luke Valenta, Benjamin VanderSloot, Eric Wustrow, Santiago Zanella-Béguelin, and Paul Zimmermann, *Imperfect Forward Secrecy: How Diffie-Hellman Fails in Practice*, 22nd ACM Conference on Computer and Communications Security (CCS '15), October 2015.

[83] CISA, "Detecting Post-Compromise Threat Activity in Microsoft Cloud Environments," Alert Code TA14-353A, updated April 15, 2021a.

[84] Peter Kornbluh and Carlos Osorio, *The CIA's 'Minerva' Secret*, Briefing Book #696, National Security Archive, February 11, 2020.

- An encryption algorithm for mobile data, GPRS GEA-1, had a flaw that was suspected to be an intentional backdoor.[85]
- Reports allege that the NSA placed a backdoor in the Dual_EC_DRBG pseudorandom number generator.[86]
- Curve25519 is an elliptic curve used in elliptic curve cryptography. It gained popularity due to concerns that the NSA had allegedly intentionally chosen values in more standardized curves that enabled them to break encryption.[87]
- Prime numbers in the Diffie-Helman algorithm could be backdoored, allowing the NSA to decrypt vast amounts of data.[88]
- The TETRA radio communications encryption standard, used by police forces across the world, had an export version with a backdoor that limited the key size to 32 bits; this backdoor had existed since 1995, but it was only publicly disclosed in 2023.[89]
- The Data Encryption Standard key size was chosen in consultation with the NSA. Allegedly, this was done to ensure that the NSA would be able to break the encryption by a brute-force attack.[90]
- Because of its potentially abusable structure, it has been hypothesized that the Russian cryptographic hash standard has a backdoor.[91]

These examples focus on cryptography and access control, but backdoors can also be intentionally introduced in software to enable code execution, overlapping with attack vectors in the Running Unauthorized Code category.

Code Vulnerabilities (in the Access Control System)

Most access control systems are either software systems or have significant software components. In addition to the more specialized ways of undermining such systems described above, an attacker could undermine them by finding code vulnerabilities and then subverting their behavior (without actually dealing with their cryptographic or core functionality at all). A major category of code vulnerabilities that undermine access control systems on a regular basis are **privilege escalation vulnerabilities**.

- Common ways to overcome such systems without engaging with their actual core security include editing them to save the passwords or keys legitimate users use to log in, degenerating their random keys to be predictable or enumerable, or configuring them to continue to function identically in all user-facing

[85] Christof Beierle, Patrick Derbez, Gregor Leander, Gaëtan Leurent, Håvard Raddum, Yann Rotella, David Rupprecht, and Lukas Stennes, "Cryptanalysis of the GPRS Encryption Algorithms GEA-1 and GEA-2," Paper 2021/819, *Eurocrypt 2021*, International Association for Cryptologic Research, 2021; Bruce Schneier, "Intentional Flaw in GPRS Encryption Algorithm GEA-1," *Schneier on Security*, blog post, June 17, 2021.

[86] Nadiya Kostyuk and Susan Landau, "Dueling over Dual_EC_DRGB: The Consequences of Corrupting a Cryptographic Standardization Process," *Harvard Law School National Security Journal*, Vol. 13, June 7, 2022.

[87] Daniel J. Bernstein and Tanja Lange, "SafeCurves: Choosing Safe Curves for Elliptic-Curve Cryptography," webpage, December 1, 2014.

[88] Dan Goodin, "NSA Could Put Undetectable 'Trapdoors' in Millions of Crypto Keys," *Ars Technica*, October 11, 2016.

[89] Kim Zetter, "Code Kept Secret for Years Reveals Its Flaw—a Backdoor," *Wired*, July 24, 2023.

[90] Bruce Schneier, "The Legacy of DES," *Schneier on Security*, blog post, October 6, 2004.

[91] Léo Perrin, "Partitions in the S-Box of Streebog and Kuznyechik," Paper 2019/092, *FSE 2019*, International Association for Cryptologic Research, 2019.

aspects but use no security or weak security instead of the original security. This is common enough that it has its own OWASP entry.[92] Examples:

- A famous early example of a privilege escalation vulnerability is described in the book *The Cuckoo's Egg*, and shortly described in Houser's review of that book.[93]
- Lazar and colleagues review 269 cryptographic code vulnerabilities reported between 2011 and 2014.[94] Note that not all cryptographic code vulnerabilities are part of a vulnerability in an access control system and vice versa, but there is significant overlap.

Access to Secret Material Undermining a Protocol

An attacker can gain illegitimate access to a key or other secret that undermines the reliability of a protocol or platform—even if that secret does not belong to the victim. This might involve gaining access to the private key of a certificate authority undermining TLS, or to any record of the parameters used to generate or acquire a code- or firmware-signing key for some ubiquitous infrastructure, such as Windows, Intel, or Cisco. This attack lies at the intersection of undermining access systems and compromising existing credentials, and it is likely only available to very well-resourced attackers. Examples:

- In 2011, an attacker gained access to the root certificate authority Comodo Cybersecurity by intruding on a reseller account, which gave the attacker the ability to issue bogus certificates.[95]
- That same year, root certificate authority DigiNotar was also breached. By similarly producing illegitimate certificates, the attack exposed the Google accounts of hundreds of thousands of Iranian users.[96]
- In 2012, Microsoft was attacked by a group that took advantage of a vulnerable hash function to create a bogus certificate authority, which was then used to spread the Flame malware.[97]
- The Content Scramble System (CSS) is a digital rights management and encryption system to protect DVDs from being copied. This system was irreversibly rendered useless by the accidental exposure of its decryption key.[98]

Bypassing Primary Security System Altogether

Attackers may not need to cope with the systems in place for security, authentication, or controlling access if they can find an alternative path (not top of mind to the security team) to the information they seek (in our context, the model weights). Examples of how such alternative paths could come about include the following attack vectors.

[92] OWASP, "A02:2021—Cryptographic Failures," webpage, undated-a.

[93] Greg Houser, "*The Cuckoo's Egg* & How it Relates to Cybersecurity," *Exida*, blog post, February 2, 2023.

[94] David Lazar, Haogang Chen, Xi Wang, and Nickolai Zeldovich, *Why Does Cryptographic Software Fail? A Case Study and Open Problems*, APSys'14, Association of Computer Machinery, June 25–26, 2014.

[95] Comodo Cybersecurity, "Update 31-MAR-2011," webpage, March 31, 2011.

[96] Josephine Wolff, "How a 2011 Hack You've Never Heard of Changed the Internet's Infrastructure," *Slate*, December 21, 2016.

[97] Dan Goodin, "Flame Malware Wielded Rare 'Collision' Crypto Attack Against Microsoft," *Ars Technica*, June 5, 2012.

[98] Andy Petrizio, "Why the DVD Hack Was a Cinch," *Wired*, November 2, 1999.

Incorrect Configuration or Security Policy Implementation

Configuration errors may unintentionally enable access. Similarly, but separately, a security policy may be established but not actually implemented. For instance, a database is accidentally left accessible even though it contains sensitive information. Examples:

- Of the more than 20,000 data breaches reported in the United States and made publicly available by government entities, more than 11 percent involved an unintentional disclosure of information by the organization itself.[99]
- In 2017, nearly 23,000 MongoDB databases were breached and ransomed because of a common misconfiguration that left the database open without password protection.[100]
- One GitHub catalog of AWS customer security incidents contains dozens of examples of unintended or illegitimate access to sensitive cloud data, many of which are due to misconfiguration.[101]
- Especially harmful instances of misconfigured AWS databases include S3 buckets containing sensitive airport and airline data,[102] as well as consumer credit data.[103]
- National Aeronautics and Space Administration (NASA) employee data were exposed due to a misconfigured Jira deployment.[104]
- In 2023, Microsoft accidentally publicly revealed 38TB of confidential data, including employees' workstation disk backups with secrets, keys, passwords, and over 30,000 internal Teams messages.[105] This was due to an Azure access token that accidentally provided permissions to more data than intended.
- In 2019, a misconfigured firewall for Capital One servers allowed unauthorized access to bank account information for 80,000 Capital One users.[106]

Additional (Less Secure) Copies of Sensitive Data

Additional copies of the weights (or other sensitive materials) may not be monitored or secured as thoroughly as the "main copy." The copy might be intentionally created (e.g., backup copies or local copies used by research teams), but the security team is unaware of its existence or has not invested in adequately securing it. Alternatively, there may be additional copies that are accidentally retained (such as training checkpoints and other files left over from training, or a copy some researcher has made that was meant to be temporary), and no one is aware that they exist. Examples:

[99] Privacy Rights Clearinghouse, "Data Breach Chronology," webpage, undated.

[100] Catalin Cimpanu, "Hacker Ransoms 23k MongoDB Databases and Threatens to Contact GDPR Authorities," *ZDNet*, July 1, 2020.

[101] Rami McCarthy (ramimac), "Background (aws-customer-security-incidents)," GitHub, undated.

[102] Nathan Eddy, "Cloud Misconfig Exposes 3TB of Sensitive Airport Data in Amazon S3 Bucket: 'Lives at Stake,'" *Dark Reading*, July 6, 2022; Claudia Glover, "Pegasus Airline Breach Sees 6.5TB of Data Left in Unsecured AWS Bucket," *TechMonitor30*, August 17, 2022.

[103] "Data on 123 Million US Households Exposed Due to Misconfigured AWS S3 Bucket," *Trend Micro*, December 20, 2017.

[104] Keumars Afifi-Sabet, "NASA Employee Data Exposed for at Least Three Weeks Due to Misconfigured Web App," *IT Pro*, January 14, 2019.

[105] "Microsoft AI Researchers Accidentally Expose 38 Terabytes of Confidential Data," *Hacker News*, September 19, 2023.

[106] Zev Brodskey, "The Capital One Data Breach: How Crisis Could Have Been Averted," Perimeter 81, July 31, 2019.

- A U.S. Department of Veterans Affairs employee had copies of records containing personal data on 17.5 million veterans and active-duty military personnel on his laptop, which was stolen.[107]
- Uber had backups of databases with user data in a nonproduction environment, which was less secure and exploited by attackers.[108]
- There are multiple examples of companies failing to properly destroy copies of data on hard drives that they then disposed of.[109]
- In 2017, an employee of London Heathrow Airport dropped an unencrypted USB device on the street, which contained a copy of confidential security measures, including security patrols, CCTV locations, and plans for protecting the Queen.[110]

Alternative (Less Secure) Authentication or Access Schemes

Many systems have nonstandard authentication or access pathways. Although there is no inherent reason for these to be less secure than the primary means of authentication, they are often more neglected and less secure. Examples of such authentication paths include an authentication protocol for when users have forgotten their password (or lost their security key), if there is a need for urgent access without proper authorization (common in codebases with multiparty authorization, see below), or when an administrator is accessing another user's account.

Searching for such opportunities requires doing so systematically and meticulously; thus, the time investment is nontrivial. However, beyond that (if the relevant mistakes are made), it should be fairly easy. Examples:

- While authentication is becoming increasingly strictm using methods such as multifactor authentication, alternative legacy authentication interfaces may remain insecure. For example, attackers compromised Citrix using an IMAP-based password-spraying campaign.[111]
- Interfaces intended for use by more technical personnel rather than the broader public can be subject to less scrutiny. For example, a hacker identified and published over half a million trivial Telnet credentials of Internet of Things (IoT) devices.[112]
- Similarly, hundreds of organizations, including Tesla, left their Kubernetes administration consoles accessible over the internet without any password protection.[113]
- Reportedly, millions of devices remain vulnerable to attacks due to open SMB and Telnet ports.[114]

---

[107] John Files, "V.A. Laptop Is Recovered, Its Data Intact," *New York Times*, June 30, 2006; Jaikumar Vijayan, "One Year Later: Five Lessons Learned from the VA Data Breach," *Computer World*, June 1, 2007.

[108] Neil Chilson, "A Lesson from Uber: Secure Your Non-Production Software Environments," Federal Trade Commission, blog post, April 12, 2018.

[109] James Kilkelly, "When the Billion Dollar Hard Drive Grows Legs," Manufacturing.net, August 31, 2015.

[110] "Heathrow Probe After 'Security Files Found on USB Stick,'" BBC News, October 29, 2017.

[111] Lindsey O'Donnell, "Threatlist: IMAP-Based Attacks Compromising Accounts at 'Unprecedented Scale,'" *Threat Post*, March 14, 2019.

[112] Silviu Stahie, "Over 500,000 Credentials for Telnet Exposed IoT Devices and Servers Leaked Online," *Bitdefender*, January 20, 2020.

[113] RedLock CSI Team, "Lessons from the Cryptojacking Attack at Tesla," RedLock, February 20, 2018.

[114] Ionut Arghire, "Millions of Devices Remain Exposed via SMB, Telnet Ports: Rapid7," *Security Week*, June 15, 2017.

AI-Specific Attack Vectors

Some attacks target the AI infrastructure itself and are only relevant for AI systems. Because this is a relatively new field, real-world examples are still scarce, but this should not lead to a false sense of security. The fact that this field is nascent likely implies that AI infrastructure is *more* vulnerable than other types of infrastructure rather than less. We should also expect that novel attack vectors will arise in the future.

The first subset of AI-specific attack vectors are ones aimed at **code execution**. Adversaries targeting the unique aspects of the ML supply chain can find significant opportunities for exploiting the large and flexible attack surface of ML systems to run malicious code. This is especially concerning due to the fast-paced development that often characterizes ML systems, leading to compromises in security and the use of unaudited firmware dependencies. These lead to a core infrastructure that is less secure than the standard most software engineers are accustomed to—for example, with high-severity issues continuing to be reported in GPU drivers at a high rate and no public bug bounty programs available.[115] As a result, even tools that are considered core ML development infrastructure and provided by security-aware and trusted industry organizations cannot be assumed to be secure.

Discovering Existing Vulnerabilities in the Machine Learning Stack

Adversaries may exploit preexisting vulnerabilities in the ML software and hardware stack (including the full supply chain) to execute malicious code or otherwise undermine the security of the models or the broader system. This is an AI-specific subset of running unauthorized code and discovering existing vulnerabilities in the supply chain; however, executing it may require less expertise on the vulnerability side (due to the current lax state of ML infrastructure) and more ML expertise. This attack vector includes using the AI model to trigger vulnerabilities in non-ML systems—for example, if the model output is not properly sanitized and can be used to execute malicious code. Examples:

- ShellTorch was a code execution exploit on TorchServe, a PyTorch model server.[116]
- An AI model could be induced to generate an output that would lead to unintended code execution, similar to the code execution vulnerability in Apache Struts reported in 2017 due to improper sanitization of error messages.[117]
- In January 2024, a GPU vulnerability dubbed LeftoverLocals was revealed,[118] allowing a malicious process on a local system to extract private GPU memory, which might include model weights or response data.
- Also in January 2024, four different critical vulnerabilities were found in the MLFlow platform.[119]
- In December 2023, a remote code execution vulnerability was revealed in Hugging-Face's Transformers library due to missing input restrictions.[120]

[115] NVIDIA, "Product Security, webpage, undated; NIST, "CVE-2023-7018 Detail," webpage, National Vulnerability Database, last modified December 29, 2023d.

[116] Oligo, "ShellTorch," webpage, undated.

[117] Fred Bals, "CVE-2017-5638: The Apache Struts Vulnerability Explained," Synopsys, blog post, September 13, 2017.

[118] Sead Fadilpašić, "PyTorch Hit by Severe Security Compromise," *TechRadar*, January 3, 2023.

[119] Shweta Sharma, "Frequent Critical Flaws Open MLFlow Users to Imminent Threats," *CSO*, January 18, 2024.

[120] NIST, 2023d.

- Popular critical ML infrastructure, such as CUDA, TensorFlow, and PyTorch, can have sprawling dependencies or be too flexible to be reliably secure.[121] Consequently, such platforms are unlikely to be secure.

Intentional Machine Learning Supply Chain Compromise

Adversaries with sufficient access or influence may undermine the ML supply chain. This can include compromising GPU hardware, data annotations, elements of the ML software stack, or the model itself. This is an AI-specific variation on the broader category of Supply Chain Attacks, but executing it may require less expertise on the vulnerability side (given the current lax state of ML infrastructure) and more ML expertise. Examples:

- PyTorch's prerelease Linux packages, PyTorch-nightly, were compromised via a malicious binary on PyPi.[122] The torchtriton dependency in the PyTorch-nightly build was replaced with a malicious package, leading to the extraction of sensitive data from infected systems.
- Google Colab's Jupyter Notebooks, often used for ML research, were manipulated to execute malicious code.[123] Shared Jupyter Notebooks on Google Colab, containing malicious code, could lead to unauthorized Google Drive access and data theft when executed by unsuspecting users.
- Several vulnerabilities in the production AI framework Ray—used by many leading AI companies—allowed attackers to compromise models, gain access to sensitive data, and steal compute for cryptocurrency mining.[124]

An additional category of attacker goal is **model manipulation**: manipulating the behavior or outputs of the model in ways that were not intended by its developers. Although the terminal goal of model integrity is outside the scope of this report, manipulating the execution of an AI model could be used as a step in a broader attempt to steal the weights if the model's correctness has security implications (e.g., identification of prompt injection attempts, classification of malicious packets on the network).

Prompt-Triggered Code Execution

This consists of bypassing restrictions of a model that is able to execute code by specially crafting prompts to manipulate the model's behavior (known as *prompt injection*). In the broader ML context (not specific to LLMs), crafting prompts relates closely to the concept of adversarial examples. Note that some prompt injection attacks may be used to execute code (such as the MathGPT example below); this subset overlaps with code execution attacks. Examples:

- The MathGPT application, which used GPT-3 to convert user queries into Python code, was found to be vulnerable to prompt injection attacks.[125] The attacker manipulated the GPT-3 model to generate code that led to unauthorized access to the application's environment variables and the ability to execute non-terminating code.

[121] TensorFlow, "Using TensorFlow Securely," GitHub, undated.

[122] MITRE, "Compromised PyTorch Dependency Chain," incident date of December 25, 2022b.

[123] MITRE, "Arbitrary Code Execution with Google Colab," incident date of July 2022a.

[124] Avi Lumelsky, Guy Kaplan, and Gal Ebaz, "ShadowRay: First Known Attack Campaign Targeting AI Workloads Actively Exploited in the Wild," Oligo, March 26, 2024.

[125] MITRE, "Achieving Code Execution in MathGPT via Prompt Injection," incident date of January 28, 2023.

- The NVIDIA AI Red Team discovered a prompt injection vulnerability in LangChain (a common LLM framework) that allows code execution.[126]

The final set of attack vectors aim at **model derivation**: leaking information about the internal workings of the model by using many model queries. This set of attacks most directly provides access to the model weights (or a functional equivalent).

Model Extraction

The answer to each query of the model bears information on the weights that helped produce it. At least in theory (and to some extent, in practice; see below), one could infer the weights from examining enough query responses. In the non-AI examples below, parameter extraction was done even to cryptographic components built especially to prevent this. Current AI architectures were not built to prevent this kind of attack. However, the sheer number of weights can introduce an operational challenge to extracting all the information (and especially doing so without being caught).

Experts with whom we spoke varied widely in how plausible they thought such an attack was, ranging from very easy to obviously infeasible. Clearly, more research is needed on this front. Another important question that may have significant implications for the feasibility of such an attack is whether extracting only a portion of the weights will be useful for a malicious attacker. Examples:

- Researchers were able to successfully extract the embedding matrices of various OpenAI models using typical API access.[127]
- Microsoft's open-source Counterfit tool can, among other things, extract weights through interaction with models (called "model inversion" in the package terminology).[128]
- Because this field is fairly young and rapidly evolving, it is valuable to glean insights from parameter extraction in other contexts. COMP-128-1 was the original algorithm used for authentication and key generation for all mobile communications in the Global System for Mobile Communications (GSM) standard.[129] Despite being designed to avoid this, the key used by this algorithm could be extracted through repeated applications.

Model Distillation

Adversaries can replicate the functionality of a private ML model by making repeated queries to its inference API. This process, known as *model distillation*, involves using the target model's inferences to train a new model that mimics the original's behavior. While similar to extracting model weights, distillation bypasses the complexity of directly accessing numerous weights. This technique is evolving, with novel approaches emerging, such as learning from explanation traces and attacks on major AI chatbots.[130] Examples:

[126] Rich Harang, "Securing LLM Systems Against Prompt Injection," NVIDIA, blog post, August 3, 2023.

[127] Nicholas Carlini, Daniel Paleka, Krishnamurthy Dvijotham, Thomas Steinke, Jonathan Hayase, A. Feder Cooper, Katherine Lee, Matthew Jagielski, Milad Nasr, Arthur Conmy, et al., "Stealing Part of a Production Language Model." arXiv, arXiv:2403.06634, March 11, 2024.

[128] Microsoft Azure (azure), "counterfit," GitHub, undated.

[129] Billy Brumley, *A3/A8 & COMP128*, T-79.514 Special Course on Cryptology, Helsinki University of Technology, November 11. 2004.

[130] Subhabrata Mukherjee, Arindam Mitra, Ganesh Jawahar, Sahaj Agarwal, Hamid Palangi, and Ahmed Awadallah, "Orca: Progressive Learning from Complex Explanation Traces of GPT-4," *arXiv* preprint 2306.02707, June 5, 2023; Will Knight, "A

- Microsoft's Counterfit tool enables "functional extraction" by interacting with models to replicate their behavior, providing a practical example of model distillation.[131]
- One study demonstrates the feasibility of model extraction attacks.[132] These attacks replicate the functionality of ML models with high fidelity using black-box access. The study highlights vulnerabilities in ML-as-a-service systems, in which confidential models are exposed via public query interfaces.

Nontrivial Access to Data or Networks

Malicious actors may gain access to information or internal networks or services that they are not expected to access. The attack vectors in this category may comprise the full attack (e.g., they give the attacker access to the weights), or they may be one step in a multistep attack (e.g., providing access to a network in which the attacker can exploit a vulnerability and then continue to explore the network, or providing a way of exfiltrating the weights after the attacker has already successfully injected code that can access them). Nontrivial access can be used both to penetrate a sensitive system and to exfiltrate information collected.

Digital Access to Air-Gapped Networks

Penetrating and exfiltrating information from a system that is not ethernet-connected but that has lots of communications (e.g., via USB sticks) is more challenging than doing so for a fully internet-connected system, but it still has significant precedent. The only unique requirement in this scenario (relative to an internet-connected device) is reaching the computer—either remotely (e.g., running code from an internet-connected device that will infect a USB stick or similar device that interacts with the air-gapped device) or through in-person intervention (see "Physical Access to Systems," below). Once the computer has been reached, running illegitimate code is broadly similar to doing so on internet-connected devices: Something needs to tell that device to run it. This is usually achieved either by tricking a user into running an executable (a common way of achieving code execution via phishing, for example) or by finding a vulnerability that leads to code execution even without user interaction (as is common in more-advanced remote code execution schemes).

More-advanced operations may manipulate insiders into changing network configurations to enable a network link where one did not previously exist. For instance, they could deliberately disrupt a critical organization service, expecting that in its urgent need to address the issue, the organization will bypass certain defenses—for example, connecting debugging devices to a system intended to remain isolated from external networks. They could also use social engineering to achieve this.

Many Secret and Top Secret classified networks perform automated exfiltration testing (see more details in the description of SL4 in Appendix B) to determine whether their air-gapped networks communicate with the internet. One expert noted that for such networks, unintended communication events occurred about 300 times per year. In addition to the approaches described above, a capable adversary can exploit such unplanned events as well. Examples:

New Attack Impacts Major AI Chatbots—and No One Knows How to Stop It," *Wired*, August 1, 2023.

[131] Microsoft Azure (azure), undated.

[132] Florian Tramèr, Fan Zhang, Ari Juels, Michael K. Reiter, and Thomas Ristenpart, *Stealing Machine Learning Models via Prediction APIs*, 25th USENIX Security Symposium, August 10–12, 2016.

- An ESET report identifies 17 malware frameworks that used USB devices to access air-gapped networks across 15 years, all suspected to be associated with state actors.[133]
- ESET's summary report provides interesting details about malware programs that target air-gapped networks, including that many operate for many years before being detected, that almost half are automatically executed when the USB is inserted using a software vulnerability, and that almost half enable command and response (and not just unidirectional exfiltration).
- An easier task is to communicate across an air gap once malware has already been installed inside an isolated network (e.g., via physical access, as described in the next category). Examples of this include the badBIOS malware,[134] Ramsay,[135] and others.[136]
- Exfiltration from air-gapped networks typically happens over USB drives; however, there are a number of known covert channels:
  - The CASPER attack uses the internal speakers inside a computer to transmit information using ultrasound frequency modulation.[137]
  - SATA cables can be used to broadcast information over 6GHz radio.[138]
  - A memory bus can be used to transmit a Wi-Fi signal over 2.4 GHz.[139]
  - There are many other examples in Mordechai Guri's publications and a survey of air-gap attacks.[140]

Side-Channel Attacks (Including Through Leaked Emanations; i.e., TEMPEST Attacks)

Electronic information can often be detected or inferred using various leaking emanations, including conducted electromagnetic emissions (e.g., electricity usage), optical emissions, acoustic sounds, radiated electromagnetic emissions, radio signals, caches (especially across cloud instances), and differential fault analysis. Such attacks (or specific subcategories of them) are known in different circles as *side-channel attacks*, *TEMPEST attacks*, *Van Eck phreaking*, *emanation monitoring*, or *compromising emanations*. Leaking the full model weights directly through a side-channel attack is considered by many to be infeasible because of the size of the weights and the low throughput of information exfiltration in side-channel attacks. Even so, side-channel attacks can be used effectively to undermine the security of the model weights in other ways—for example, by exfiltrating a key that enables decrypting model weights or authenticating to receive access to the weights. Examples:

---

[133] Alexis Dorais-Joncas and Facundo Muñoz, *Jumping the Air Gap: 15 Years of Nation-State Effort*, ESET, December 2021a.

[134] Dan Goodin, "Meet 'badBIOS,' the Mysterious Mac and PC Malware That Jumps Airgaps," *Ars Technica*, October 31, 2013.

[135] Ignacio Sanmillan, "Ramsay: A Cyber-Espionage Toolkit Tailored for Air-Gapped Networks," We Live Security, May 13, 2020.

[136] Mordechai Guri, Gabi Kedma, Assaf Kachlon, and Yuval Elovici, "AirHopper: Bridging the Air-Gap Between Isolated Networks and Mobile Phones Using Radio Frequencies," *arXiv* preprint arXiv:1411.0237, November 2, 2014.

[137] Bill Toulas, "CASPER Attack Steals Data Using Air-Gapped Computer's Internal Speaker," *Bleeping Computer*, March 12, 2023.

[138] Physics arXiv Blog, "Hack Forces Air-Gapped Computers to Transmit Their Own Secret Data," *Discover*, July 29, 2022.

[139] Tom Spring, "Air-Gap Attack Turns Memory Modules into Wi-Fi Radios," *Threat Post*, December 17, 2020.

[140] For Guri's research, see Mordechai Guri, "Air-Gap Research," webpage, undated; and Andy Greenberg, "Mind the Gap: This Researcher Steals Data with Noise, Light, and Magnets," *Wired*, February 7, 2018a. The survey is described in Jangyong Park, Jaehoon Yoo, Jaehyun Yu, Jiho Lee, and Jae Seung Song, "A Survey on Air-Gap Attacks: Fundamentals, Transport Means, Attack Scenarios and Challenges," *Sensors*, Vol. 23, No. 6, 2023.

- While U.S. awareness of TEMPEST attacks began during World War 2,[141] followed by decades of research and deployment of TEMPEST defenses and exploitation, they were first discussed in an unclassified paper in 1985.[142]
- Among many other examples,[143] the CIA analyzed an Egyptian cipher machine for emanations and then detected them through microphones in an office below the Egyptian embassy.
- The book *Information Warfare* includes examples of TEMPEST attacks in industrial espionage.[144]
- The Navy's 1988 *Automated Information Systems Security Guidelines* manual notes that "foreign governments continually engage in attacks against U.S. secure communications and information processing facilities for the sole purpose of exploiting CE [compromising emanations]."[145]
- Over time, costs for such attacks have fallen. One example (of many) in a paper by Genken and colleagues is the extraction of decryption keys via electrical emanations from outside a room with equipment that costs $3,000.[146] Publications by Eran Tromer and Yossi Oren provide many other examples.[147]
- Researchers discovered a side-channel vulnerability in a hardware optimization of Apple Silicon CPUs that allows extraction of secret keys.[148]
- A 2023 paper by Harrison, Toreini, and Mehrnezhad shows that (at least in some circumstances) passwords can be identified with high probability by listening to keyboard keystrokes through a mobile phone's microphone.[149] In practice, this means an attacker could steal corporate passwords (or other information) by successfully running malware on an employee's personal phone. The referenced paper is the most recent high-profile one (at time of writing in mid-2024), but there are thousands of such papers published in the literature.
- AI accelerators such as GPUs are relatively nascent and, as a consequence, are more vulnerable. They may leak weight information (or cryptographic information undermining the security of the weights) by side-channel means, such as temperature, sound, power consumption, and resource sharing. It has been shown that in current multi-GPU[150] and multi-application systems,[151] information can be leaked between different components of the system.

[141] *TEMPEST: A Signal Problem*, National Security Agency, September 27, 2007.

[142] Wim van Eck, "Electromagnetic Radiation from Video Display Units: An Eavesdropping Risk?" *Computers & Security*, Vol. 4, No. 4, December 1985.

[143] David Easter, "The Impact of 'Tempest' on Anglo-American Communications Security and Intelligence, 1943–1970," *Intelligence and National Security*, Vol. 36, No. 1, 2021.

[144] Winn Schwartau, *Information Warfare*, 2nd ed., Thunder's Mouth Press, 1996.

[145] U.S. Navy, "Emanations Security," in *Automated Information Systems Security Guidelines*, 1988; archived March 30, 2008.

[146] Daniel Genkin, Lev Pachmanov, and Itamar Pipman, "ECDH Key-Extraction via Low-Bandwidth Electromagnetic Attacks on PCs," *RSA Conference Cryptographers' Track (CT-RSA) 2016*, LNCS 9610, Springer, 2016.

[147] Eran Tromer, "LEISec: Laboratory for Experimental Information Security," webpage, undated; Yossi Oren, "Oren Lab—Implementation Security and Side-Channel Attacks: Publications," webpage, undated.

[148] GoFetch, homepage, undated.

[149] Joshua Harrison, Ehsan Toreini, and Maryam Mehrnezhad, "A Practical Deep Learning-Based Acoustic Side Channel Attack on Keyboards," *arXiv* preprint 2308.01074, August 2, 2023.

[150] Sankha Baran Dutta, Hoda Naghibijouybari, Arjun Gupta, Nael Abu-Ghazaleh, Andres Marquez, and Kevin Barker, "Spy in the GPU-Box: Covert and Side Channel Attacks on Multi-GPU Systems," *Proceedings of the 50th Annual International Symposium on Computer Architecture*, 2023.

[151] H. Naghibijouybari, A. Neupane, Z. Qian, and N. Abu-Ghazaleh, "Side Channel Attacks on GPUs," *IEEE Transactions on Dependable and Secure Computing*, Vol. 18, No. 4, July–August 2021.

- Side-channel attacks against AI models are still nascent as well. In the future, one might expect to see more specialized and advanced attacks, similar to those used to recover RSA private keys.[152]
- Ross Anderson's *Security Engineering* has a wealth of additional real-world examples of side-channel attacks.[153]

Eavesdropping and Wiretaps

An attacker can take other actions to collect sensitive information in a continuous manner, including sending individuals to physically listen to conversations (with or without the assistance of specialized devices); planting microphones, cameras, or other automated devices; tapping wires (phone lines, ethernet cables, internet backbone, etc.) to covertly access communicated data; and other approaches. Like many of the attacks described in this report, this attack vector could comprise the full attack (e.g., capturing the model weights in transit) or be one component in a broader effort (a tool to extract information to be used for extortion, passwords, or keys to breach defense systems, etc.). Examples:

- Eavesdropping can be greatly enhanced using some more-advanced techniques, such as laser microphones used to listen to conversations as far as 1,600 feet away by the Soviet Union as early as 1947.[154] Today, a rudimentary laser microphone can be built with materials costing a few dollars.[155]
- A recent vulnerability in a Zoom feature exemplifies a purely digital form of eavesdropping, while also illustrating how one attack can quickly facilitate another (also enabling remote code execution).[156]

Unauthorized Physical Access to Systems

This category can be thought of as a subcategory of the previous one. In the majority of cases, physical access to a system can be translated into meaningful access to sensitive information on the system. Even if the underlying data are encrypted or the information is protected against physical attacks, intelligent attackers can install devices or software that would provide them access later on—for example (physical or digital) keyloggers that would save and/or send passwords or keys when a legitimate user accesses the system, or software that would run and extract information directly whenever a legitimate user accesses it. Therefore, encryption alone (and other similar defenses) is not sufficient to prevent an attacker with physical access from accessing sensitive data.

To make this concept slightly more concrete, here are two specific ways to translate one-time physical access into longer-term control over a device (for standard PCs), at different levels of effort/sophistication:

- Even a simple attacker can buy a $80 USB drive that injects keyboard strokes and leave the drive in or near the office.[157] This would provide code execution, though not necessarily exfiltration.

[152] Daniel Genkin, Adi Shamir, and Eran Tromer, "RSA Key Extraction via Low-Bandwidth Acoustic Cryptanalysis," *CRYPTO 2014*, part I, LNCS 8616, Springer, 2014.

[153] Ross Anderson, *Security Engineering: A Guide to Building Dependable Distributed Systems*, 3rd ed., Wiley, 2020.

[154] Albert Glinsky, *Theremin: Ether Music and Espionage*, University of Illinois Press, 2005.

[155] Rick Osgood, "Fast Hacks #6—Laser Spy Microphone," video, July 16, 2023.

[156] "Zoom ZTP & AudioCodes Phones Flaws Uncovered, Exposing Users to Eavesdropping," *Hacker News*, August 12, 2023.

[157] Hak5, "USB Rubber Ducky," webpage, undated-b.

- For a slightly more effective attack, there are USB cables concealing a chip that provides remote control and communications; these can be purchased for as little as $180.[158]

This type of attack is often considered more risky: The possibility of being caught is higher, and physical crime is often more heavily penalized than virtual crime. However, the extent to which this approach is actually risky depends critically on how complex the operation is. If a device is easy to access or a location is easy to break into (whether this is in the office, at home, etc.—see options below), capable adversaries can anonymously hire local criminals to achieve access with nearly no risk to themselves. Of the more than 20,000 data breaches reported in the United States and made publicly available by government entities, almost 12 percent involved some physical element (though the definition of *physical element* in the source may differ from ours).[159]

An attacker can achieve physical access to a system in several ways.

Direct Physical Access to Sensitive Systems

If no exceptional security practices are in place, an attacker can often directly access physical servers (or other devices). This may involve social engineering ("I was just in there but forgot my . . ."), breaking in (with a spectrum of possible levels of sophistication), impersonating those with legitimate justification to enter (e.g., employee, client), and other creative means. It may involve achieving authorized access even without impersonation—maintenance, cleaning services, support, etc. Note that the maintenance or services do not need to be related to the sensitive system; any service that has access to a room with sensitive information or a device that can access sensitive information will suffice.

Another common point of access is waste disposal. Electronic storage devices (or other materials containing sensitive materials) that are thrown away without being wiped, shredded, or otherwise destroyed give an attacker an easy and low-risk opportunity. Examples:

- In 2019, *Yahoo News* reported that, as part of the Stuxnet operation, a mechanic allegedly plugged a USB device either directly into the control systems or into the device of an engineer.[160]
- Attackers posing as overnight janitors installed keyloggers on computers at Sumitomo Mitsui Bank in London in an attempt to steal hundreds of millions of pounds.[161]

Malicious Placement of Portable Devices

A common and easy-to-execute strategy is to leave devices (most commonly USB flash drives, but also hard drives, computer peripherals, and other devices) laying around the target facility (in the office, in the parking lot, in the elevator). These devices could be marked in some way to give them credibility (the logo of the organization, the name of an employee, some other label implying that it is related to the organization), though they need not be marked. It is very common for people to find these devices and plug them into a computer as a way of understanding whether they have something important on them or what should be done with them.

[158] Sean Gallagher, "Playing NSA, Hardware Hackers Build USB Cable That Can Attack," *Ars Technica*, January 20, 2015; Hak5, undated-a.

[159] Privacy Rights Clearinghouse, undated.

[160] Kim Zetter and Huib Modderkolk, "Revealed: How a Secret Dutch Mole Aided the U.S.-Israeli Stuxnet Cyberattack on Iran," *Yahoo News*, September 2, 2019.

[161] Peter Warren and Michael Streeter, "Mission Impossible at the Sumitomo Bank," *The Register*, April 13, 2005.

Although awareness of the risks of plugging in an untrusted USB device is slowly increasing, so are the capabilities of attackers. We already noted USB cables with hidden chips that provide advanced access, and very few people are careful about which USB cable they use.[162]

This attack vector is fairly easy, and it is likely to work unless there is a clear policy of not plugging in untrusted devices, or (more reliably) devices in the system are configured to either not communicate with (non-approved) external devices or to report when this happens. Examples:

- There are several examples of USB sticks or similar devices infecting U.S. nuclear power plants.[163]
- Agent.btz, the malware that infected the U.S. Department of Defense in 2008 and led to the creation of U.S. Cyber Command, came from a USB device that was found in a parking lot on a base in the Middle East.[164]
- There are reported cases of "evil twin" attacks, in which an attacker accesses a victim's device via a fraudulent Wi-Fi network.[165]

Physical Access to Devices in Other Locations

If the premises are well secured, attackers could try to access organization devices in other locations. They may break into an employee's home, steal a laptop or connect something to it in a public space, ask an employee to make an urgent call from their mobile device and install something on it, etc.

There are also more "advanced" ways of getting access to people's devices, even if they do not leave them unattended in public spaces. States often access people's devices at the airport (often under the guise of a security inspection), and capable actors more broadly (including states) will have the infrastructure in place to access people's devices when left in hotel rooms.

An even more sophisticated option is receiving permission to enter sensitive premises (including employee homes) under false pretenses, including work relationships (such as cleaners, plumbers, etc., who are let into the house) or other relationships (guests, one-night stands, etc.). It is worth noting that drive encryption is useful in this context; however, even encrypted, locked, and sometimes turned-off devices might be vulnerable, depending on circumstances. Ways to overcome drive encryption include installing cameras in the hotel room or home to collect a password and using that to decrypt a copied encrypted hard drive, or copying (usually unencrypted) RAM memory. Examples:

- Sidd Bikkannavar, a U.S. employee of NASA's Jet Propulsion Laboratory, was detained and pressured into allowing his NASA-issued phone to be searched by the U.S. Customs and Border Patrol after returning from a trip to Chile.[166]
- News outlets reported that the contents of former Secretary of Commerce Carlos Gutierrez's laptop were believed to be copied during a state trip to China in late 2007.[167]

---

[162] Hak5, undated-a.

[163] See Section 3.4, "Malware Attacks to US Nuclear Power Plants," in Bıçakcı, 2015.

[164] Alexander Gostev, "Agent.btz: A Source of Inspiration?" Kaspersky SecureList, March 12, 2014.

[165] Kaspersky, "Evil Twin Attacks and How to Prevent Them," webpage, undated-a.

[166] Loren Grush, "A US-Born NASA Scientist Was Detained at the Border Until He Unlocked His Phone," *The Verge*, February 12, 2017.

[167] "Did Chinese Hack Cabinet Secretary's Laptop?" NBC News, May 29, 2008.

Evasion of Physical Access Control Systems

Many physical security systems critically rely on digital systems (e.g., cameras, alert systems, digital locks). If necessary, these can be compromised or disabled through other means, including finding security vulnerabilities. This vector partially overlaps with Running Unauthorized Code: It is a similar activity, only directed toward a specific category of systems. Examples:

- Researchers from cybersecurity firm F-Secure were able to exploit a flaw in digital lock company Assa Abloy's Vision system to create copies of hotel master keys.[168]
- A vulnerability in Onity hotel locks, revealed at Black Hat, allowed an attacker to break into dozens of hotel rooms using a homemade device. Although the fix required installing new hardware in every affected lock, Onity refused to pay for replacement circuit boards for months after the reveal; many hotel locks may still be vulnerable.[169]
- Security researchers used vulnerabilities in hotel keycards using older radio frequency identification (RFID) technology to forge the keycards, demonstrating that millions of hotel rooms are vulnerable.[170]
- A group of "hacktivists" gained access to hundreds of thousands of Verkada security cameras, giving them access to live feeds, video archives, and in some cases internal network access to the affected companies. Although it was not done in this instance, such access could also be used to understand security camera blind spots or even disable security cameras during a break in.[171]

Penetration of Physical Hardware Security

In high-security contexts, some devices will have physical hardware protections (e.g., HSMs, smartcards, and other secure cryptoprocessors—see more details in Appendix B). In these cases, in addition to physically reaching the device, an attacker would need to overcome physical hurdles to access information on it. This is difficult but possible. Capable actors have a large toolset for gaining access, including drilling, electromagnetic probing, acid dissolvent, unusual clock signals, freezing, and timing/power attacks. It is an unfortunate (and not very well-known) characteristic of digital hardware (including specialized security hardware) that it will do many unexpected things when fuzzed effectively.

This type of attack can be challenging, and probably only large organizations that perform physical attacks at scale develop the capacity and expertise to do this. However, it is not beyond the capabilities of skilled state actors. Examples:

- Cold boot attacks, in which data remnants stored in RAM (like encryption keys) can be extracted by freezing the physical RAM with liquid nitrogen and then rebooting the computer using specially designed software[172]

[168] Jenny Southan, "A Faked Master Key Gives Hackers Access to Millions of Hotel Rooms," *Wired*, April 25, 2018.

[169] Andy Greenberg and Lisk Feng, "The Hotel Room Hacker," *Wired*, August 2017.

[170] Unsaflok, homepage, undated.

[171] Chaim Gartenberg, "Security Startup Verkada Hack Exposes 150,000 Security Cameras in Tesla Factories, Jails, and More," *The Verge*, March 9, 2021.

[172] J. Alex Halderman, Seth D. Schoen, Nadia Heninger, William Clarkson, William Paul, Joseph A. Calandrino, Ariel J. Feldman, Jacob Appelbaum, and Edward W. Felten, *Lest We Remember: Cold Boot Attacks on Encryption Keys*, Princeton University Center for Information Technology Policy, 2008.

- Integrated circuit (IC) decapsulating, in which an attacker physically removes the IC cover, allowing the chip to be reverse engineered or code to be copied[173]
- Clock glitching and crowbar glitching, attacks that disrupt the device clock and voltage, respectively, lead to erroneous behavior, such as skipped instructions.[174]
- Laser fault injection is a method that fires a laser at a transistor during its operation, disrupting its behavior and potentially allowing access to restricted files.[175]

Armed Break-In

This vector is a variation on direct physical access, but with much more capable agents who are both better trained in gaining access and have a broader toolset at their disposal—including drugging security personnel and using violence if necessary. It may (aim to) be a covert operation or not. This approach is significantly more costly than the simple direct physical access option.

It is also a significantly higher risk: The malicious organization directing the operation will not want the agents on the ground to be arrested, so things can escalate much more quickly. In addition, if the attack is detected, it is easier to trace it to its source: The people on the ground who carry out the attack require significant training and resources, so they are often more clearly connected to the adversary. Example:

- Attackers broke into a data facility in Chicago, incapacitated the guard, and stole thousands of dollars in server equipment. This was the fourth time the facility was broken into in two years.[176]

Military Takeover

At the extreme end of physical attacks, an attacker might use military force to take over a facility. While this option has costs (see below), it enables a wide range of effective tools, including overcoming physical barriers much more easily or the ability to execute much more complex operations within the facility.

This option is available only to state actors and exceptionally large criminal organizations. But for such organizations, it is not too difficult, and they have people trained to attack much more challenging targets. However, this option entails enormous cost and will likely provoke significant retaliation. Examples:

- On the day of the 2021 Myanmar coup, the military directed armed soldiers to break into data centers nationwide.[177] Employees who resisted were held at gunpoint.
- If specific frontier models gain strategic national security importance, one could start seeing foreign military attacks on them as well, as has already occurred with a nuclear power plant in Ukraine.[178]

---

[173] Kyle Orland, "MAME Devs Are Cracking Open Arcade Chips to Get Around DRM," *Ars Technica*, July 25, 2017.

[174] NewAE Technology, "Glitching," webpage, undated.

[175] Ledger Academy, "Episode 3—Laser Fault Attacks," webpage, updated June 4, 2023.

[176] Dan Goodin, "Masked Thieves Storm into Chicago Colocation (Again!)," *The Register*, November 2, 2007.

[177] Fanny Potkin and Poppy McPherson, "Insight: How Myanmar's Military Moved in on the Telecoms Sector to Spy on Citizens," Reuters, May 18, 2021.

[178] Edith M. Lederer, "UN Nuclear Chief: Ukraine Nuclear Plant Is 'Out of Control,'" Associated Press, August 3, 2022.

Supply Chain Attacks

Supply chain attacks involve intentionally undermining the security of supply chains, either surreptitiously (e.g., exploiting the openness of open-source development, breaching the security of a supply chain supplier) or through other types of influence (e.g., governments may require this by law or have secret agreements with supplier companies). This category is similar to the Intentional Backdoors in Algorithms, Protocols, or Products (in the Access Control System) subcategory above, though that category focuses on the cryptographic or algorithmic undermining of access control systems, whereas this category focuses on gaining access to systems and networks (usually for code execution).

Supply chain attacks are not conducted directly against the organization or the code it has written but against third-party suppliers that the organization utilizes. Even an organization that has perfectly secure coding practices may be vulnerable to an immense variety of vulnerabilities based on the enormous infrastructure it relies on.

The attack surface of supply chain attacks is vast. We highlight a few examples of what supply chain attacks can target (and that require separate mitigation efforts).

Services and Equipment the Organization Uses

This includes computers, paraphernalia, GPUs and ML-specific hardware, electric wiring in the walls, USB devices that might be connected, security keys, printers and office equipment, chip-embedded code (e.g., Intel's AMT vulnerability, Silent Bob Is Silent),[179] operating systems, code repositories, Continuous Integration and Continuous Deployment infrastructure, compiled artifact repositories, developer sandboxes, and staging systems, monitoring and instrumentation, and other proximate systems that are directly involved in building and maintaining research or deployed production models, ML infrastructure, documentation software, chat software, workspace software (e.g., Slack), video conference software, and generally installed apps (browsers, note apps, etc.). Unless very aggressive policies are put in place, the attack surface remains immense. Examples:

- Operation Rubicon, in which the CIA and German intelligence covertly purchased Swiss cypher machine manufacturer CryptoAG, giving them access to classified information from dozens of countries over many decades[180]
- The SolarWinds hack, which effected a wide-ranging supply chain attack that, in turn, affected companies including Microsoft, Intel, and Cisco—themselves companies that many AI organizations rely on as suppliers of key hardware and software across many of their operations.[181] It also affected many U.S. government departments, including the Pentagon and CISA, highlighting the potential effectiveness of a supply chain attack even against organizations that prioritize security.
- The Target breach that exposed the financial information of 110 million customers used the chain's HVAC system to penetrate the network.[182] Reportedly many HVAC systems have such vulnerabilities,

[179] Dan Goodin, "Intel Patches Remote Hijacking Vulnerability That Lurked in Chips for 7 Years," *Ars Technica*, May 1, 2017.

[180] Melina Dobson, Jason Dymydiuk, and Sarah Mainwaring, "Operation Rubicon: The Most Successful Intelligence Heist of the 20th Century," Warwick Knowledge Center, undated.

[181] Dina Temple-Raston, "A 'Worst Nightmare' Cyberattack: The Untold Story of the SolarWinds Hack," *All Things Considered*, NPR, April 16, 2021.

[182] "Top 7 Cybersecurity Threats: #3 Supply Chain Attacks," *MxD*, October 11, 2022.

and experts in the national security community advise that low-tech infrastructure (beyond HVACs specifically) is rife with such vulnerabilities.[183]

- China's supply chain attacks on computer hardware allegedly provide access to the servers of top U.S. companies and government suppliers.[184]
- Hundreds of malwares by multiple attackers used code-signing keys stolen from Microsoft to attack Windows computers directly or attack software suppliers as part of advanced supply chain attacks in recent years.[185]
- Millions of servers inside data centers imperiled by firmware vulnerabilities were revealed in a breach of hardware maker Gigabyte[186]
- The ChaosDB attacks on the Microsoft cloud platform, Azure, allowed the remote takeover of client accounts.[187]
- In the recent Storm-0558 attack on Outlook, attackers got hold of a cryptographic key that allowed them to create fake authentication tokens, which they used to access the accounts of dozens of organizations.[188]
- The 2022 Okta breach (one of several) enabled the Lapsus$ group to access 366 companies through the identity and access management provider.[189] The breach of Okta itself was through a third-party vendor with access to its systems.
- Alleged efforts by China to utilize Huawei, ZTE, and other companies to infiltrate or disrupt critical U.S. operations.[190]

Code and Infrastructure Incorporated into the Codebase

Some code and infrastructure are not developed by the organization; they are developed by third parties and then incorporated into the organization's codebase. These third-party components include code packages, libraries, images, tooling, development infrastructure, and more.

Given the enormous reliance of the modern software world on open-source infrastructure, third-party code comprises a very significant portion (and often the majority) of most organizations' code. Examples:

- According to Synopsys's Open Source Security and Risk Analysis report,[191] 48 percent of open-source codebases include at least one known high-risk vulnerability.

---

[183] Jaikumar Vijayan, "Target Attack Shows Danger of Remotely Accessible HVAC Systems," *Computer World*, February 7, 2014.

[184] Micah Lee and Henrik Moltke, "Everybody Does It: The Messy Truth About Infiltrating Computer Supply Chains," *The Intercept*, January 24, 2019.

[185] Goodin, 2023d.

[186] Dan Goodin, "Firmware Vulnerabilities in Millions of Computers Could Give Hackers Superuser Status," *Ars Technica*, July 20, 2023c.

[187] Wiz, "ChaosDB," webpage, undated.

[188] Andy Greenberg, "How a Cloud Flaw Gave Chinese Spies a Key to Microsoft's Kingdom," *Wired*, July 12, 2023.

[189] Zack Whittaker, "Okta Says Hundreds of Companies Impacted by Security Breach," *TechCrunch*, March 23, 2022.

[190] Katie Bo Lillis, "CNN Exclusive: FBI Investigation Determined Chinese-Made Huawei Equipment Could Disrupt US Nuclear Arsenal Communications," CNN, July 25, 2022.

[191] Synopsys, "[2023] Open Source Security and Risk Analysis Report," webpage, undated.

- In 2024, it was discovered that a common Linux SSH utility included a malicious backdoor, unknowingly incorporated by an open-source project maintainer.[192]
- Package managers are frequent targets of malicious packages. In 2022, a malicious package was discovered within PyPi that swapped the user's cryptocurrency wallet address with the attacker's address if the former was ever copied onto the clipboard.[193] In 2024, PyPi suspended account and project creation briefly after being inundated with mass uploading of malicious packages using typosquatting.[194]
- In 2018, malicious code that harvested Bitcoins and wallet keys was added as a dependency of the widely used event-streams package of NPM and was not discovered for nearly two months, resulting in it being included in 8 million downloads.[195]
- A flaw in Atlassian's Jira software allowed attackers to access unused signup tokens, giving them access to Jira Service Management instances.[196]
- The Log4Shell vulnerability existed in the ubiquitous Java logging framework Log4J for close to a decade and was exploited by cybercriminals before it was disclosed.[197] The vulnerability also remained exploitable across many systems long after it was patched, and to this day, 11 percent of Java codebases remain vulnerable to it.[198]

Vendors with Access to Information

A separate attack vector is vendors who have access to sensitive data or systems; if the vendors are compromised (even on a completely separate network), an organization's data can be compromised. This includes vendors with consistent and ubiquitous access to the organization's data (e.g., GitHub, security tools that have access to the code), as well as vendors used for specific use cases (e.g., IBM experts who help to configure servers, FinOps experts who automatically manage cloud spending on the organization's behalf, a security-managed service provider who helps to handle ongoing data loss prevention alerts). Examples:

- *Yahoo News* reported that the Stuxnet attack involved setting up multiple front companies just to get a single mechanic into Natanz to deliver a payload via USB and attacked five vendors as prerequisites for the attack.[199]
- OPM was breached by an APT based in China, affecting tens of millions of SF-86 records. Attackers used a credential they had stolen from a subcontractor to gain access to OPM.[200]

---

[192] Andres Freund, "Backdoor in Upstream xz/liblzma Leading to SSH Server Compromise," memorandum to OSS Security, March 29, 2024.

[193] Dan Goodin, "Latest Attack on PyPI Users Shows Crooks Are Only Getting Better," *Ars Technica*, February 14, 2023b.

[194] Yehuda Gelb, Jossef Harush Kadouri, and Tzachi Zornshtain, "PyPi Is Under Attack: Project Creation and User Registration Suspended," Checkmarx, blog post, March 28, 2024.

[195] Danny Grander and Liran Tal, "A Post-Mortem of the Malicious Event-Stream Backdoor," *Snyk Blog*, December 6, 2018.

[196] Ravie Lakshmanan, "Atlassian's Jira Service Management Found Vulnerable to Critical Vulnerability," *Hacker News*, February 3, 2023a.

[197] CISA, "Mitigating Log4Shell and Other Log4j-Related Vulnerabilities," Advisory, Alert Code AA21-356A, updated December 23, 2021b; Andreas Berger, "What is Log4Shell? The Log4j Vulnerability Explained (and What to Do About It)," Dynatrace, blog post, updated June 1, 2023.

[198] Lily Hay Newman, "A Year Later, That Brutal Log4j Vulnerability Is Still Lurking," *Wired*, December 10, 2022; Synopsys, undated.

[199] Zetter and Modderkolk, 2019; Jon Fingas, "Stuxnet Worm Entered Iran's Nuclear Facilities Through Hacked Suppliers," *Engadget*, November 13, 2014.

[200] Koerner, 2016.

Human Intelligence

Some attacks involve influencing people rather than influencing technological systems (though, as always, the boundaries are fuzzy). Like other categories, human intelligence may be used as a component in a broader attack (e.g., to gain access to the network, then continue to use other tools).

Despite being much less openly discussed than digital cyberattacks, human-based infiltrations occur at large scales (and, obviously, predate the more modern and technical means of espionage). Regardless of the specific interaction used to persuade or force an individual to cooperate, such attacks will often begin by identifying individuals with access who are vulnerable or influential in some way. One common entry point is through professional relationships, with the attacker misrepresenting their identity and engaging in an enticing business or academic proposition or relationship. Other times it includes engaging through positive interactions with a member of the relevant gender (depending on the target). Both are well known to occur in professional conferences, social and leisure venues frequented by target groups, and online. Details on some common methodologies used by intelligence agencies and how to be more aware and protected against them are described in a booklet published by the Defense Security Service and National Counterintelligence and Security Center.[201]

There is an important fact that makes all approaches described below both more feasible and less risky than many people assume. People without experience in human intelligence often imagine that if an adversary wants an employee of some organization to assist in undermining their organization's security, the adversary needs to pay, threaten, or otherwise convince the employee to do so. This would make success more challenging, because many employees may refuse to cooperate due to their loyalty to the organization. It would also pose risk to the adversary because reporting such an incident would implicate them.

In practice, the adversary need not tell the employee who they are and can easily pretend to be any other actor. They can also misrepresent their goals. This not only significantly reduces the risk to the adversary but also enables them to tailor their identity and goals to the ideology of the employee. A story that aligns with the employee's own goals significantly improves the chances of success. If the employee believes that their organization should be more transparent than it is, the story will amplify that element. If the employee believes AI progress should be advancing more slowly, the adversary can craft a scheme that (allegedly) achieves that goal.

It is also worth noting that while this category focuses on human intelligence and intentional insider threats, a large portion of insider risk results from nonmalicious insiders. The threats that arise because of mistakes, poor cyber hygiene, and so on are not described in this section but in other sections depending on context—including under social engineering, exploitation of exposed credentials, incorrect configuration or security policy implementation, and others.

Here are several approaches.

Bribes and Cooperation

People can be paid or persuaded to take actions on behalf of an adversary. This could include getting them to do something "as a favor" or simply cooperating with people who already have an agenda (e.g., disgruntled employees seeking revenge, ideological employees serving a cause). This agenda or ideology can be nurtured and encouraged by the adversary. Note that, in accordance with the principles described above, individuals do not have to be bribed to (knowingly) do something that is wrong, nor do they need to perceive this as a

[201] Defense Security Service and National Counterintelligence and Security Center, "Counterintelligence: Best Practices for Cleared Industry," undated.

bribe. They can often be fooled into thinking that what they are doing is legitimate (e.g., "we're paying you to try out this great new demo product"). Examples:

- Ana Montes was a Cuban spy within the U.S. intelligence community for well over a decade, recruited based on her views against U.S. policy toward Central America.[202]
- Robert Hanssen, "the most damaging spy in Bureau history," claimed that his motivation was purely financial (although those who knew him suggest additional motivators).[203]
- In 2023, two Navy sailors were arrested for transmitting sensitive military information to the People's Republic of China in exchange for bribes,[204] one of the sailors pleaded.

Extortion

Instead of providing benefits, an adversary can threaten a person with physical harm, legal harm, financial harm, and more. State actors might have *substantial* leverage (e.g., putting people in jail), but even fairly noncapable actors can have significant leverage: Even minor criminals can find out where a person lives and threaten to stab them. Leverage can also be applied to friends and family of the target.

Adversaries will often generate their own leverage for extortion—for example, convincing someone to take some small illegitimate action (e.g., just copy this file) under the pretense this is not a big deal, then clarify that it was a huge deal and threaten to turn them in unless they continue to cooperate. This could range from simple tactics, such as persuading a local criminal to offer someone $1,000 to do something, to more extreme tactics, such as systematically arresting or kidnapping and torturing family members. The severity of the tactic influences both the difficulty and the amount of risk such a strategy would entail. Example:

- Russia was accused of forcing Ukrainian intelligence officers to spy for Russia by threatening to kill their families.[205]
- China has been reported to intimidate, threaten, and coerce exiles and individuals across the world at unparalleled scale.[206]

Candidate Placement

In a more sophisticated variation, capable adversaries may employ and train people before they join the organization and help them become strong candidates for sensitive roles. This is a much more costly (and less opportunistic) methodology, but once deployed it can be much more effective in ensuring reliable and extensive access. Note that this does not require training people for the specific organization: Many organizations filter candidates based on similar criteria, so only limited adjustments are needed per company. Thus, actors who use this attack vector at scale (e.g., capable state actors) can draw on existing agents who were trained long before the specific organization was set as the target. Examples:

[202] FBI, "Ana Montes: Cuban Spy," webpage, undated-a.

[203] "What Made the American Turncoat Tick?" CNN, May 10, 2002.

[204] "2 Navy Sailors Arrested, Accused of Providing China with Information," CBS News, updated August 3, 2023.

[205] Joe Barnes, "Russia Forces Ukrainians to Become Spies by Threatening to Kill Their Families," *The Telegraph*, February 6, 2024.

[206] Freedom House, *China: Transnational Repression Origin Country Case Study*, 2021.

- Marcus Klingberg, deputy head of the Institute of Biological Research in Israel, who passed information regarding Israeli chemical and biological weapons research to the Soviet Union[207]
- Kim Philby, MI5 executive who was a Russian agent[208]
- Hanjuan Jin, software engineer at Motorola Inc., accused of sending confidential materials to China.[209]

Organizational Leverage Attacks

At the more strategic or institutional level, the organization itself can be coerced or fooled into playing into the adversary's hand. An adversary can build financial or legal leverage over an organization—for example, through investments or grants that appear innocent initially but are then used to force the organization into giving access. This can be done without the organization realizing that it is now compromised. For example, an investor can promote a collaboration with a different institution that includes providing them legitimate access (e.g., using their software). Example:

- The U.S. Department of Justice accused Huawei of using its legitimate business with U.S. companies to steal trade secrets.[210]

Organizationally Approved Access

Like organizational leverage attacks, these attacks can be implemented with the known cooperation of some or all of the executive leaders of the organization. The organization can be paid to provide access to sensitive materials, often covertly. Example:

- Carnivore was a wiretap tool developed and used by the FBI to monitor internet usage during the late 1990s and early 2000s. Its use was permissible by court authorization.[211]

---

[207] William Grimes, "Marcus Klingberg, Highest-Ranking Soviet Spy Caught in Israel, Dies at 97," *New York Times*, December 3, 2015.

[208] "Harold 'Kim' Philby and the Cambridge Three," Nova Online, undated.

[209] FBI, Chicago Division, "Suburban Chicago Woman Sentenced to Four Years in Prison for Stealing Motorola Trade Secrets Before Boarding Plane to China," August 29, 2012.

[210] U.S. Department of Justice, "Chinese Telecommunications Conglomerate Huawei and Subsidiaries Charged in Racketeering Conspiracy and Conspiracy to Steal Trade Secrets," press release, February 13, 2020.

[211] FBI, "Robert Hanssen," webpage, undated-b.

APPENDIX B

Detailed Benchmark Systems for Security Levels

In this appendix, we provide full descriptions of the security measures listed in the summary tables of the security level benchmarks in Chapter 6. Within each benchmark (both here and in the summary tables in Chapter 6), we organize the security measures by custom categories and subcategories adapted to the specific context of securing AI model weights and aimed at increasing readability. However, to assist in understanding how the measures relate to existing frameworks, we also include the relevant NIST Cybersecurity Framework version 1.1 function and category for each category in short form (e.g., ID.AM).[1]

Security Level 1 (SL1)

Weight Security

Weight Storage (PR.DS)

- **Sensitive data remain internal.** Employees are expected to keep sensitive information on organization devices, networks, and cloud instances and are trusted to do so.
- **Weight encryption (best effort).** Weights are encrypted with at least 128-bit encryption, both in storage and in transport through an untrusted network (e.g., the internet).

Physical Security (PR.DS)

- **Data centers of cloud providers.** Weights are stored primarily (but not necessarily exclusively) in cloud instances of standard cloud providers. Those copies are in data centers that have security measures (including security guards) to prevent unauthorized access to the servers.

Access Control (PR.AC)

- **Access control for sensitive assets.** There is an access control system for the codebase, main data storage system, the weights, and other materials designated as sensitive by the security team. Access is provided only to relevant subgroups of employees.
- **Access log or audit trail.** Access to the weights (or the storage unit the weights are in) is logged, including which account accessed them and when.

Security of Network and Other (Nonweight) Sensitive Assets

Software (PR.MA)

- **Moderately frequent software update management and compliance monitoring.** The IT team applies software updates on corporate devices within two weeks of release or within a month in case of substantial difficulty or risk (e.g., downtime-sensitive production systems).

[1] NIST, undated.

Access, Permissions, and Credentials (PR.AC)

- **Least privilege principle.** Users have the lowest privilege necessary, based on a small number of predefined roles, such as engineering, Human Resources, and IT.
- **Restrictions on device and account sharing.** Each user is assigned a specific device and account for exclusive use; sharing devices or accounts is prohibited.
- **Password best practices.** There are minimum length (often 8 characters) and complexity requirements (often a number and nonuniform capitalization) for passwords; password sharing under any circumstances is prohibited.
- **Multifactor authentication.** All devices and accounts with sensitive access require multifactor authentication. The factors used are usually a password and either text messages or authentication apps.
- **Single Sign-On (SSO).** The number of attack surfaces is minimized by limiting the number of distinct ways to authenticate with corporate systems.
- **Backup and recovery tools.** Backup and recovery tools are available to enable IT to quickly remove compromised devices or accounts.
- **Commercial identity and access management (IAM) tools.** Software is able to report such cloud access issues as lockouts, suspicious login attempts, expired passwords, and other events that need attention.
- **Zero Trust architecture.** The organization adheres to at least the standards in the "Traditional" level of CISA's Zero Trust Maturity Model.[2]

Hardware (ID.AM)

- **Modern device architectures that establish root of trust and block malicious code execution.** All devices with sensitive access have modern architectures (e.g., from the past four years) that establish a root of trust (e.g., Trusted Platform Modules [TPMs]) and provide other industry-standard tools to make malicious code execution more difficult.
- **CPU anti-exploitation features.** All modern CPUs have features aimed to mitigate some memory corruption vulnerabilities.

Supply Chain (ID.SC)

- **The reputability of software is reviewed before incorporation.** Before integrating software that is central to the organization's operations, the team reviews the software's security reputation to ensure that the software has been tried, tested, and incorporated by trusted organizations. The team also compares the reputation of the software with potential alternatives to identify the most reputable option.

Security Tooling (PR.PT)

- **Modern authentication infrastructure.** There is a reasonably up-to-date authentication infrastructure (e.g., using current standard Identity Providers [IdP], authentication services, or authentication packages) for controlling authentication before access to sensitive resources.
  - For most organizations whose investment in security is closer to SL1 and SL2, we believe that relying on unmodified existing infrastructure is better than developing anything proprietary, because attempting the latter is more likely to introduce unintended vulnerabilities than fix known ones.
  - At the level of investment of SL3 or above, more proprietary investment is warranted, built on well-established infrastructure.
- **Commercial network security solutions.** The organization has implemented, for example, a firewall that includes application-level inspection.

---

[2] CISA, 2023a.

- **Commercial endpoint security solutions.** The organization has implemented, for example, antivirus software.
- **Reliance on standard security infrastructure (depending on circumstances).** As with many aspects of security, the extent to which an organization should rely on existing external solutions and platforms versus proprietary internal ones will depend on the organization. Small and mid-scale organizations are probably better served by relying on large and trusted security platforms; extremely well-resourced organizations (with security needs that fit higher security levels) are more likely to benefit from adding customized and tailored solutions to existing security infrastructure.

Configuration Management (PR.IP)

- **Enforce screen locks for inactivity.** Screens lock after a short period of inactivity (e.g., 60 seconds).

Personnel Security

Awareness and Training (PR.AT)

- **Basic onboarding information security training for employees.** This includes password best practices, remote working protocols, multifactor authentication, etc.

Security Assurance and Testing

Risk and Security Assessments (ID.RA)

- **Internal reviews.** The internal security team routinely—at least once per year—looks for flaws and vulnerabilities in the system.

Security Team Capacity (ID.RM)

- **Basic incident response capabilities.** The security team (which may be a single person) is able to prepare for, detect, contain, and recover from a rudimentary breach. If such a breach happens, the team is also expected to audit the logs, record the occurrence of an event, try to identify the flaw, and close the vulnerability to prevent future access.

Maintenance (PR.MA)

- **Information security news monitoring and implementation.** The security team (which may be a single person) informally keeps up to date on information security news and best practices, and plans and implements updates or improvements to security based on them. Extra focus should be on AI-specific sources, such as the Institute of Electrical and Electronics Engineers' Conference on Secure and Trustworthy Machine Learning.[3]

[3] Institute of Electrical and Electronics Engineers, "2nd IEEE Conference on Secure and Trustworthy Machine Learning," conference homepage, University of Toronto, April 9–11, 2024.

Security Level 2 (SL2)

Implementation of Previous Security Levels

- The organization has implemented all the controls from SL1.

Weight Security

Weight Storage (PR.DS)

- **Storage location.** Weights are stored exclusively on servers, not on local devices, such as laptops. These servers have built-in access control and encryption to support the requirements below. The weights may not be copied such that they become an arbitrary file on an arbitrary file system.
- **Encryption.** Weights are encrypted in storage with at least 256-bit strength encryption. Decryption keys require authentication (e.g., generated from a password and/or other authentication device), ensuring cryptographic protection beyond just programmatic access control. All keys are secured in a key management system (e.g., Google's Keystore), ensuring that they are not stored in plaintext or in other easily exploitable formats.

Security During Transport and Use (PR.DS)

- **Encryption in transit.** Plaintext weights are not transported over public or unencrypted channels.

Physical Security (PR.DS)

- **Data centers are guarded, and only people with authorization are allowed inside.**
  - If the organization stores weights in data centers of standard cloud providers, this requirement is satisfied.
- **Visitor access is restricted and logged.**

Access Control (PR.AC)

- **Restrictions on sensitive interactions.**
  - Sensitive interactions include direct access to the weights, code that can access the weights, or any code or configuration files that affect the security of the system.
  - Only relevant employees have access to any sensitive assets. This includes the full relevant development team.
  - Interactions can be done only from an internal network (or, if done remotely, then via VPN).
  - Multifactor authentication using FIDO authentication/hardware security keys (e.g., yubikeys) is required.[4]
  - FIDO/hardware authentication must be done either for every interaction or for any action that is attempted after a short period (e.g., 60 seconds) since the user's last authentication.
  - All such interactions are recorded and can be reviewed.

Monitoring (DE.CM)

- **Logging of all sensitive interactions.** All sensitive interactions are recorded and stored in a write-but-no-edit system.
- **Regulation and monitoring of weight copies across the organization network.** There may be multiple untracked copies of the weights within the internal organization network (e.g., researchers may copy

[4] For additional advice, see CISA, "Implementing Phishing-Resistant MFA," fact sheet, October 2022.

them to a separate server to use them), but these requirements apply to all copies. There is either a mechanism to prevent the creation of copies in violation of these requirements, or such copies are regularly searched for within the network to ensure compliance.

AI Model Resilience

AI-specific security mitigations are generally nascent and fast-evolving. We expect recommendations in this space to change rapidly over the coming years. We include the items below to share the current state of the art, but we see these measures as less reliable than the majority of the measures mentioned.

Model Robustness (PR.DS)

- **Input reconstruction.** During inference, a privately known prefix (sometimes called a *model prompt* or *pre-prompt*) is added ahead of the user prompt. This prompt is changed occasionally, including if it is ever revealed. The prefix should limit the control of the attacker on the output, and other methods may be considered (such as adding a random noise to the input).
- **Adversarial training.** As part of the training process, adversarial examples generated using known attack techniques are added to the dataset.

Security of Network and Other (Nonweight) Sensitive Assets

Software (PR.MA)

- **Frequent software update management and compliance monitoring.** All software is kept reasonably up to date (e.g., personal devices must install security updates within 7 days, production systems must install security updates within 21 days); access is disallowed whenever this has not happened. Patch management tools are used to ensure that no update has been missed. The policy also applies to firmware and microcode (which are somewhere on the boundary between hardware and software).

Access, Permissions, and Credentials (PR.AC)

- **Strong password enforcement.** Example rules to be enforced include at least 60 bits of entropy, rotation every several months, barring using the name or other user characteristics as part of the password, banning sharing, etc.[5]
- **The work network is separate from the guest network**, either physically or logically (e.g., via virtual local area network [VLAN]).
- **Guest accounts disabled whenever possible.** Many organizations do not need them, and they create an opportunity to gain a backdoor.
- **Strong access management tools.** The security team keeps track of which individuals, devices, services, roles, and processes have access to which resources (e.g., via Cloud Infrastructure Entitlement Management) and frequently remove any unnecessary (or no longer necessary) permissions according to a least-privilege access policy.
- **Zero Trust architecture.** The organization adheres to at least the standards in the "Initial" level of CISA's Zero Trust Maturity Model.[6]

[5] Paul A. Grassi, James L. Fenton, Elaine M. Newton, Ray A. Perlner, Andrew R. Regenscheid, William E. Burr, Justin P. Richer, Naomi B. Lefkovitz, Jamie M. Danker, Yee-Yin Choong, Kristen K. Greene, and Mary F. Theofanos, *Digital Identity Guidelines: Authentication and Lifecycle Management*, National Institute of Standards and Technology, NIST Special Publication 800-63B, October 16, 2023.

[6] CISA, 2023a.

Hardware (ID.AM)

- **Lost or stolen devices reported.** It is very easy and blameless to report a lost or stolen device and have it remotely deactivated.
- **All network devices are visible and trackable** (to identify lockouts, computers with disks that are not encrypted, devices without multifactor authentication enabled, etc.).

Supply Chain (ID.SC)

- **Review of vendor and supplier security.** Security standards for vendors and suppliers comply with the organization's security requirements; all new vendors and suppliers are reviewed by the security team.

Security Tooling (PR.PT)

- **Disk encryption** for all corporate devices using well-known and commonly used 256-bit encryption (or better).
- **Network communications are encrypted by default.** Examples include using SSH, Hypertext Transfer Protocol Secure (HTTPS), Wi-Fi Protected Access (WPA), Server Message Block 3.0 (SMBv3), Secure Lightweight Directory Access (LDAPS), etc.
- **Email security tools.** Implementation or incorporation of email security tools (e.g., Abnormal Security).
- **Use of integrated security approaches**, such as eXtended Detection and Response (XDR).[7]

Configuration Management (PR.IP)

- **Incorporate fundamental infrastructure and policies for Security-by-Design and Security-by-Default.**[8]
- **Configuration management monitoring.** The security team monitors critical security configurations on an ongoing basis—manually, automatically, or both.

Physical Security (ID.AM)

- **Office security.** Security cameras, alarm systems, and security guards are utilized. These might be managed by the organization or the facility (e.g., the building that the office is in). ID readers owned and managed by the organization control entrance to areas considered secure.
- **Careful disposal of printed materials.** Work-related printed materials are shredded (rather than put in the trash).

Personnel Security

Awareness and Training (PR.AT)

- **Periodic mandatory information security training for all employees.** This includes sound training on social engineering schemes, correct management of security credentials, never plugging in an untrusted cable or device (e.g., USB) to their work devices, physical office security, and guidelines for reporting suspicious activity.
- **Employee training on configuration errors and their security implications.** Potential configuration errors with security implications are mapped out, and employees who are relevant to the security of weights (e.g., manage storage configuration, site reliability engineers [SREs], etc.) are given training.

---

[7] Anne Aarness, "What Is XDR? How to Choose an XDR Solution," *CrowdStrike*, April 18, 2023.

[8] For example, see CISA, *Secure-by-Design: Shifting the Balance of Cybersecurity Risk: Principles and Approaches for Secure by Design Software*, October 25, 2023b.

Filtering and Monitoring (PR.AT)

- **Installation of monitoring software for secure network access.** Monitoring software (e.g., installed root certification authorities [CAs], etc.) is installed before devices can receive access to an internal network or other sensitive assets.
- **Active drills to identify and educate noncompliant employees.** The security team actively executes phishing attempts (and other tactics available to opportunistic attackers) to identify and educate employees who are not compliant with the security policies.

Security Assurance and Testing

Red-Teaming and Penetration Testing (DE.AE)

- **Mandatory external reviews.** The system is reviewed and penetration tested by an accredited third-party organization before the system is considered secure.

Community Involvement/Reporting (DE.AE)

- **Bug-bounty and vulnerability-discovery programs.**

Software Development Process (PR.IP)

- **Secure software development standards.** The organization follows the guidelines provided in NIST's Secure Software Development Framework to identify and implement secure development approaches throughout the development cycle.[9]

Incident Response (RS.RP)

- **Protocols and funding for rapid incident response.**[10] The organization has clearly defined threat detection and internal advisory protocols and has allocated sufficient funding for a qualified incident response team.
- **Incident reporting.** Severe breaches, or suspicion thereof, are reported to law enforcement and other relevant stakeholders.

Security Team Capacity (ID.RM)

- **Constant availability of qualified personnel.** A qualified team (internal or external) that is capable of responding to security breaches is on call and available 24/7.

Maintenance (PR.MA)

- **Continuous vulnerability management and adaptation to information security developments.** The security team diligently and consistently updates, adapts, and responds to the latest information security developments and vulnerabilities. Staying informed about recent threats, the team prioritizes, manages, and patches vulnerabilities within the system, ensuring that top-priority issues are addressed promptly. While the team remains informed by externally discovered vulnerabilities and guidelines, it also proactively searches for system-specific vulnerabilities instead of relying solely on external patches.

---

[9] NIST, "Secure Software Development Framework," January 23, 2024.

[10] Paul Cichonski, Thomas Millar, Tim Grance, and Karen Scarfone, *Computer Security Incident Handling Guide*, NIST SP 800-61 Rev. 2, National Institute of Standards and Technology, August 2012.

Other Organization Policies (ID.RM)

- **Promotion of a security mindset by organization management.** The organization's management publicly backs the importance of having a security mindset.
- **Stringent remote work policies.** Use of work devices for personal activities is prohibited, restrictions to avoid sharing devices or credentials are enforced, access to organization network and resource from unprotected Wi-Fi is prohibited, etc.

Security Level 3 (SL3)

Implementation of Previous Security Levels

- The organization has implemented all the controls from SL1 and SL2.

Weight Security

Weight Storage (PR.DS)

- **Centralized and restricted management of weight storage.** Weights are stored in a small number of centrally managed locations (which may include a separate copy for training, research environment, production, etc., but employees and researchers cannot simply make an additional copy).
- **Secure cloud network (if applicable).** If weights are stored on a cloud provider, the cloud network is dedicated to sensitive data (e.g., AWS GovCloud, not standard AWS)
- **Dedicated devices for weights and weight security data.** All devices that store weights or information critical for weight security (such as encryption keys) do not share a machine with less-secure applications or uses (and specifically, do not use devices shared with other organizations, as is common in cloud providers).

Physical Security (ID.AM)

- **Data centers are guarded or locked at all times.** Floor plans and security practices ensure that every path from publicly accessible spaces to the data center machines includes at least two locked entryways or manned checkpoints at all times, comprehensively preventing access to servers without authorization. Physical security is handled by the organization or by a high-end professional security service (as opposed to by the building owner).
- **Premises are swept for intruders frequently** (e.g., hourly).
- **Premises are meticulously swept for unauthorized devices routinely** (e.g., monthly).

Permitted Interfaces (PR.AC)

- **Authorized users who interact with the weights do so only through a software interface that reduces risk of the weights being illegitimately copied.** This precludes, among other things, even authorized users gaining direct read access to any copy of the weights, because this enables trivial copying. There may be interfaces implementing different hardening approaches, but all interfaces to all copies of the weights are constrained by at least one of the approved constraints. We offer several recommendations below; however, the security team may decide some of these are not sufficiently strong or that others should be added:
  - **Interface alternative A: Use services, interfaces, and functions to access the weights that the security team has reviewed and certified as difficult to abuse for the purpose of extracting the full weights.**

    - Production inference could be supported by supporting only text as an input and only the output token(s) (or logits). This interface would be (indirectly) accessible to the broader user base.
    - Reinforcement learning and other fine-tuning efforts could be supported by enabling two interfaces. The first one implements a *forward pass*: It accepts input embedding vectors and outputs the results from the final layer of the model. The second one implements *backpropagation*: It accepts gradients for the final layer and outputs the gradients of the inputs. These interfaces would be accessible only to trusted researchers.
    - Note that the requirements for code being hard to abuse are significantly more extensive than requiring that the code does not directly copy the weights. For example, code that simply reads a single weight by index can be rerun many times to extract all weights. Code that allows running the model with an arbitrary loss function (returning the loss) can also be easily used by an engineer to extract the weights.
  - **Interface alternative B: Allow authorized users to execute arbitrary code accessing the weights in a secure environment that limits the total (aggregate) output rate (e.g., to be less than 1 percent of the size of the weights per year).**
    - For example, a server storing the weights may be set up with two interfaces: One serves as input-only, allowing authorized users to upload code and data onto the server, and another returns outputs and is limited to 400KB per hour.
    - This could support more flexible or spontaneous interactions for individual researchers (as they require significantly more flexibility but cannot read more than about 100 bits per second). Approved researchers could receive a monthly quota of 1GB of output to use as they wish. Because some frontier models are already 1TB or larger (and the sizes of frontier models are only expected to grow), even if a researcher (or multiple researchers) were to use their full quota to exfiltrate weights nonstop, the time required to do so reduces the feasibility or usefulness of such an effort. Additional development may be needed to enable data-efficient communication and help ensure that users do not unknowingly or accidentally spend their whole quota. This should be complemented by a hard limit on the number of accounts or identities that receive quota (for example 50 accounts).
    - Because 4-bit versions of models achieve decent performance, one should assume 4-bit weights when calculating model weight size and setting rate limits accordingly: Even if the model uses full precision (higher bit) weights, only 4 bits of each weight might be copied, and good performance could plausibly still be achieved. Additionally, because of the common use of mixture-of-expert architectures, it may be the case that exfiltrating even a portion of the model weights (e.g., 25 percent) might provide dangerous capabilities. It is likely that additional "efficiencies" may be discovered. Therefore, one should add a significant "buffer" when setting these limits—for example, assume the likely "effective" size of the weights is at least an order of magnitude less than their actual current size in memory, and preferably even larger discounts.
    - Relatedly, it is not clear that small percentages of the model weights (e.g., 1 percent), or any information about the weights of such size, are meaningfully useful and need to be protected. However, this is a nascent field, and this assessment may change. Similarly, if future developments lead to much smaller models gaining capabilities whose proliferation would be dangerous, the rate-limiting approach may cease to be tenable.
  - **Interface alternative C: An isolated network is used.**
    - Engineers can interact with the weights freely (using arbitrary code and extracting unlimited outputs) if they do so on a computer or network that is isolated from network connections, has all external connections (USB, Bluetooth, etc.) disabled, and has security to ensure that electronic

storage and other devices do not enter or leave the room. This appears to be a promising approach for the most flexible needs, though it comes at a higher cost and is more difficult to scale. It relies on the assumption that the inherent throughput of nondigital means (people's memory, paper, etc.) is low. It may be more difficult over time to ensure that individuals are not carrying digital technology of any form. For some additional details on what is needed to isolate a network, see the "Hardware" section in the SL4 benchmark.

    - Intentional and authorized copying of the full weights (setting up a new data center, duplicating a production model, etc.) requires per-instance approval by multiple people, including a key executive stakeholder (e.g., chief information security officer), enables a predetermined and pre-approved number of copies that are within the access control system (see above), and is monitored and recorded.
  - Note that many existing workflows, including popular software for debugging and interpretability research, do not satisfy any of the above alternatives, because they provide full weight access.
- **Any code accessing the weights minimizes attack surface, provides only simple forms of access, and uses the minimal amount of (highly trusted and well-established) external code necessary.**
- **Avoiding model interactions that bypass monitoring or constraints.** For sufficiently capable models, there should be no separate API where the limitations or monitoring described above are circumvented.

Access Control (PR.AC)

- **Protocols and policies for sensitive interactions.** All sensitive interactions (including access to the weights themselves rather than using them for inference, and any editing of the code of the weights interface system) are protected in the following ways:
  - **Access to the various permitted interfaces to the weights is stringently controlled.** While inference is (indirectly) accessible to the public, more-flexible interfaces are restricted to people with a concrete need. Output-limited arbitrary access is limited to 100 people, access to an isolated network with direct access is limited to 50 people, and the ability to make copies of the weights is limited to 20 people.
    - No third-party service has access to the weights.
    - No one has persistent access (e.g., web access to code rather than checking out code locally, permissions need to be renewed periodically).
    - Multiparty authorization. Updates to sensitive code need to be approved by at least two people. It is critical that multiparty authorization of security-relevant code cannot be bypassed—either through a system feature (e.g., force submit), alternative access (e.g., accessing files directly rather than through standard tooling), or existing bypass vulnerabilities.
    - Security review for sensitive interactions. Any sensitive actions cannot be immediately executed and require a security review by security personnel (this may introduce a significant delay).

Monitoring (DE.CM)

- **Ongoing manual monitoring of sensitive interactions.** Ongoing (randomly sampled or intelligently sampled) manual monitoring of such communications, both incoming and outgoing information.
- **Ongoing automated anomaly detection.** This includes significant increases in throughput, changes to traffic behavior and patterns, large amounts of nontext or non-human-readable LLM outputs, long connection times, large increases in the number of connections, and more.
- **Automated and manual monitoring/blocking of potentially malicious queries:**
  - queries that "jailbreak" the model
  - queries about dangerous content
  - queries that attempt to perform a breach (e.g., injections)

- queries that appear to try to apply distillation or model inversion (including anomalous amounts of queries)
- general anomaly detection—for example, by detecting unusual patterns in the number of queries, connection throughput, types of questions, model answer characteristics, and more
- Note that anomalous behavior is aggregated across a user's interactions (to be detected even if the anomalous volume is only apparent across many small sessions of a single user), and across all users for a specific time (so that the onset of an attack using many fictitious users can be detected).

Note that when malicious or harmful queries that match the above criteria are detected, they are proactively blocked. For instance, if a distillation attempt is detected, a system (external to the model itself) intervenes to indicate that the model cannot or will not provide an answer, and/or if a severe attempt is detected, it will suspend the account that generates the queries.

- **Frequent compromise assessment.**
- **Frequent integrity checks via comparison against a baseline system configuration ("gold image").**
- **A gold image copy of all software relevant to the security of the weights is maintained.** The copy is read-only, secured, and hashed, and all copies of the software in actual use are compared frequently against the gold image copy as an integrity check. A mismatch with the gold image copy could indicate a breach, but even if it is not due to a breach (but a simple mistake), it is still critical to identify and correct to avoid accidental degradation of security.
  - Such gold image strategies are also critical for a separate security goal: maintaining the integrity of the system. This is not the focus of our report, but the gold image strategy can be expanded to support the integrity of the system more broadly, not just those components of the system responsible for the confidentiality of the weights.

Standard Compliance (ID.GV)

- **Implementation of measures described by NIST SP 800-171 or equivalent.** This standard represents the requirements for contractors working with Controlled Unclassified Information (CUI) in nonfederal systems and organizations, published by NIST.[11] While we reference a standard here, the security level benchmarks do not represent a standard and should only be used for calibration between security measures and security outcomes. In cases in which a security measure parallel to or more strict than a NIST SP 800-171 requirement appears in an earlier security level benchmark, this implies that the security measure (or an equivalent alternative) is needed to protect even against the operational capacity for that earlier security level.
- **Future implementation of measures described by CMMC 2.0 Level 3.** The U.S. Department of Defense is currently defining its Cybersecurity Maturity Model Certification (CMMC) 2.0 Level 3 ("Expert") requirements for contractors to protect CUI from APTs.[12] Once the certification is finalized, we expect it to comprise a useful set of requirements for protecting against persistent nonstate actors. While we reference a standard here, the security level benchmarks still do not represent a standard and should only be used for calibration between security measures and security outcomes. Specifically, while CMMC 2.0 Level 3 requires a compliance assessment process, we do not mean to imply any compliance process here but merely reference the security measures described. In cases in which a security measure parallel to a

---

[11] Ron Ross, Victoria Pillitteri, Kelley Dempsey, Mark Riddle, and Gary Guissanie, *Protecting Controlled Unclassified Information in Nonfederal Systems and Organizations*, National Institute of Standards and Technology, NIST SP 800-171 Rev. 2, updated January 28, 2021.

[12] Chief Information Officer, U.S. Department of Defense, undated.

NIST SP 800-171 requirement or more strict appears in an earlier security level benchmark, this implies that the security measure (or an equivalent alternative) is needed to protect even against the operational capacity for that earlier security level.

AI Model Resilience

AI-specific security mitigations are generally nascent and fast-evolving. We expect recommendations in this space to change rapidly over the coming years. We include the items below to share the current state of the art, but we see these measures as less reliable than the majority of the measures mentioned.

Model Robustness (PR.DS)

- **Adversarial input detection.** A separate component (possibly an independent AI model) classifies adversarial inputs and blocks them from continued processing.

Oracle Protection (PR.DS)

- **Limitations on the number of inferences using the same credentials.** The number of inferences per second allowed is rate-limited using the same credentials (and enforced across multiple simultaneous connections). Even a limit of 100 tokens per second would likely require months to exfiltrate all the weights (and more as the size of models grows).

Security of Network and Other (Nonweight) Sensitive Assets

Software (PR.MA)

- **Very frequent software update management and compliance monitoring.** All software is kept aggressively up to date (e.g., personal devices must install critical security updates within two days, production systems must install critical security updates within seven days); access is disallowed whenever this has not happened. Patch management tools are used to ensure that no update has been missed. The policy also applies to firmware and microcode (which is somewhere on the boundary between hardware and software).

Access, Permissions, and Credentials (PR.AC)

- **802.1x authentication.** The 802.1x protocol is used to secure wired and wireless networks.
- **Zero Trust architecture.** The organization adheres to at least the standards in the "Advanced" level of CISA's Zero Trust Maturity Model.

Hardware (ID.SC)

- **Security-minded hardware sourcing.** All work devices (including servers, employee devices, and peripherals) are sourced from providers that have been reviewed and found to be relatively secure (e.g., reputable and independent, avoiding companies with allegations of espionage or undermining security).

Supply Chain (ID.SC)

- **Software inventory management.** All software installed on organization devices is tracked and managed (to ensure comprehensive implementation of policies below). Software that is known to be problematic, does not conform to the policies below, or makes comprehensive enforcement of the policies difficult is *blocklisted* (active uninstallation or active prevention of code execution).
- **Supply chain security is commensurate with the organization's security.** Any supply chain provider whose employees, code, or services could access sensitive organization materials (including but not lim-

ited to model weights) must be validated to achieve security commensurate with that of the organization.

Security Tooling (PR.PT)

- **Enforcement of security policies through code rather than manual compliance.** Whenever feasible, all security policies are implemented by code, not by relying on people being instructed to act in a specific way (e.g., links in emails from outside the organization are disabled, USB ports are disabled for non-approved devices). A great example of the importance of this principle is the millions of emails sent to Mali (.ml) instead of the U.S. military (.mil).[13]
- **Security policy enforcement for network access across devices.** Compliance with security policies—e.g., required security/monitoring software, hardware encryption, limitations on installed software—is a prerequisite for access to the organization network (nonwork devices, such as personal phones, either cannot connect or a similar policy is enforced).
  - This applies to *all* devices, including printers, keyboards, thermostats, etc. All the inventory is constantly monitored, and no devices hold credentials that are not required for their function.

Personnel Security

Awareness and Training (PR.AT)

- **Employee awareness of weight interaction monitoring.** Employees are made aware that interactions with the weights are monitored (to create deterrence among nonaligned employees).
- **Security training for employees** (not necessarily only those with access).
  - Security training to avoid phishing and other compromise attempts (e.g., how to identify if an employee is being approached)
  - Clear guidelines for reporting any suspicious activities (preferably including active red-teaming attempts by the security team to attack to check whether employees report them)
- **Security risk reporting program.** Any security risk can be easily reported to leadership and it will be investigated.

Filtering and Monitoring (PR.AT)

- **Insider threat program,** the implementation of which is guided by CISA's *Insider Threat Mitigation Guide* and the National Insider Threat Task Force's *Insider Threat Program Maturity Framework*.[14] Specifically:
  - Background checks are conducted for all employees. Employees with access to the weights or any sensitive systems go through extensive screening every six months.
  - Employees are trained to report suspicious activity and given the information and tools to do so effectively (what is suspicious, what do you do if you suspect an insider threat). According to a U.S. Department of Defense report,[15] in almost 40 percent of convicted employee exfiltration cases, people noticed concerning behavior or changes in behavior before the perpetrators were arrested.

---

[13] Bernd Debusmann, Jr., "Typo Sends Millions of US Military Emails to Russian Ally Mali," BBC News, July 17, 2023.

[14] CISA, *Insider Threat Mitigation Guide*, November 2020b; National Insider Threat Task Force, *Insider Threat Program Maturity Framework*, 2018.

[15] Stephanie L. Jaros, Katlin J. Rhyner, Shannen M. McGrath, and Erik R. Gregory, *The Resource Exfiltration Project: Findings from DoD Cases, 1985–2017*, Defense Personnel and Security Research Center, Office of People Analytics, 2019.

- Employees are aware of the severe consequences (for them and more broadly) of a data breach. This may include informing employees that the organization will aggressively prosecute any tampering with sensitive files or systems, whether intentionally or due to negligence. Having employees sign a statement related to this appears to increase awareness and compliance.
- We also recommend reviewing Carnegie Mellon University's *Common Sense Guide to Mitigating Insider Threats* and the "Understanding Adversaries" chapter in *Building Secure and Reliable Systems*, both of which provide recommendations for addressing insider threats.[16]

Security Assurance and Testing

Red-Teaming and Penetration Testing (DE.AE)

- **Ongoing penetration testing.** The security team continuously performs high-quality penetration testing, with special attention given to the interface to the weights.
- **Penetration testing of physical access and facility security.**
- **Advanced red-teaming:**
  - **Elite external team.** The third-party team is extremely capable and can effectively simulate an actor with the relevant capabilities. Such teams will commonly (although not necessarily) include people with significant experience in offensive-focused cybersecurity organizations (such as the Microsoft Security Response Center or Google's Project Zero), knowledge of APT techniques and operating system internals, and a proven record of finding and exploiting vulnerabilities.
  - **Substantial funding.** The third-party team receives significant funding to pull off complicated attacks.
  - **Access to design and code.** The third-party team is given access to the system design and code (whitebox red-teaming[17]).
  - **Testing insider threats.** The third-party team is given employee credentials (to be able to test insider threats).
  - **Expanded access.** In general, the third-party team can get any access or information they request if it enables testing a different part of the system.
  - **Attention to the weights and authentication.** Special attention is given to the security of the interface of the weights and its authentication system.

Risk and Security Assessments (ID.RA)

- **Keeping a risk register.** The system's security risks are identified and logged.

Threat Detection and Response (RS.RP)

- **Placement of effective honeypots.** Honeypots are systems or resources designed to be attractive to potential attackers, which help notify the security team of illegitimate access. For example, a fake copy of the weights could be placed on a server that no employee has access to but is not properly secured, triggering an alert if the files are ever accessed. Recognizing that no system is ever completely secure, honeypots increase the chance that detection, mitigation, and response occur before irreversible harm is

[16] Computer Emergency Response Team, National Insider Threat Center, *Common Sense Guide to Mitigating Insider Threats*, Carnegie Mellon University, 2022; Heather Adkins, Betsy Beyer, Paul Blankinship, Piotr Lewandowski, Ana Oprea, and Adam Stubblefield, *Building Secure and Reliable Systems: Best Practices for Designing, Implementing, and Maintaining Systems*, O'Reilly, 2020, Chapter 2: Understanding Adversaries.

[17] Mark Nicholls, "Types of Penetration Testing: Black Box, White Box & Grey Box," Redscan, December 10, 2023.

done (or at least earlier in the process). Most employees neither know about nor are able to review what honeypots are in use. (The use of honeypots is controversial; many experts believe them to be ineffective.)

Security Team Capacity (ID.RM)

- **General increased capacity (compared with SL2).** The team has a capacity of at least two dozen people or 5 percent of organization headcount, whichever is larger.
- **Concrete experience with APTs.** The security team includes at least one person with real-world, hands-on experience with capable-state actor-level APT (either in offense or defense).
- **Leveraging diverse security experience from leading organizations.** The team has information security experience at multiple leading organizations.

Other Organization Policies (ID.RM)

- **Two independent security layers.** The security team maintains, as part of its defense-in-depth strategy,[18] a concept of independent security layers—layers of defense that are each believed to remain secure even if all other layers are completely undermined. In reviews of the security of the system (including red-teaming described above), each security layer is tested independently, and a failure of one is considered to be a failure of the system's security. Independent security layers are important to make the security system robust against occasional access to zero-days or novel approaches.

Security Level 4 (SL4)

Implementation of Previous Security Levels

- The organization has implemented all the controls from SL1–SL3.

Weight Security

Weight Storage (PR.DS)

- **Isolation of weight storage.** Weights are stored only in *protected setups* that are physically separate from the external world (e.g., a separate server not used for other purposes), with specifically defined and managed connections to the external world that conform the policies described in the current section (weight storage) and the "Permitted Interfaces" section below.
  - There can be multiple protected setups, each containing a copy of the weights. Different protected setups may span different scales, with their "line of defense" in different places. In some cases, the protected setup is a single device, and all constraints below apply to any connection it has; in other cases, the protected setup is a whole data center, and the constraints apply to communications between the data center and the external world (including the AI organization's offices themselves).
  - We suspect that this (including the requirements below) is not feasible with standard service from cloud providers, though it may be achievable through a more boutique service.
- **Weight storage setup is protected against eavesdropping and the simplest of TEMPEST attacks.** For example, devices that store or access weights do not share a room with untrusted devices, cannot be seen through windows, are not near thin walls shared with an untrusted space, etc. Note that this level does

[18] Cloudflare, "What Is Defense in Depth? Layered Security," webpage, undated-a.

not require full protection from TEMPEST attacks (e.g., a Sensitive Compartmented Information Facility [SCIF]-like solution; see more details and requirements in the description of SL5). It also implies that any cloud solution without at least a separate room for the organization's hardware is not sufficiently secure.

- **Hardware-enforced limits on output rate.** All continuous output communications between the protected setup and the external world (e.g., all network connections) are rate-limited at the hardware level (e.g., through the use of cables with limited bandwidth). The aggregate output rate across all connections must not exceed a certain portion of the size of the weights per year (e.g., 5 percent). More research is needed for more accurately setting this threshold. Note that the requirement in SL3 applied only to a subset of interfaces with the weights, whereas this requirement applies to all of them.
- **Reduced communication capabilities.** All feasible steps are taken to harden devices and reduce any communication capabilities not required for the permitted interfaces (e.g., as defined in SL3) in the protected setup (e.g., remove camera, speakers, audio board, Wi-Fi module, and antennae; remove or epoxy USB ports; disable any type of autorun; refuse all external connections). This makes physical access more difficult and so is important even if there are no relevant devices in the room.

Security During Transport and Use (PR.DS)

- **Confidential computing (when available).** Confidential computing, a technique for securing data while in use by ensuring a hardware-based trusted execution environment (TEE), is used to secure the weights. The use of confidential computing in GPUs is still nascent and may not be "production-ready," but we include it because of an overwhelming consensus by experts regarding its importance.
  - Because confidential computing is a broader paradigm that can be used for a variety of security goals, it is important that it is implemented to provide the model weights with sufficient confidentiality. Specifically, one should ensure the following:
    - The TEE does include protections against physical attacks (current implementations of confidential computing in GPUs do not[19]).
    - Model weights are encrypted by a key generated within the TEE and stored only within the TEE.
    - The TEE will only run prespecified and audited signed code. That code decrypts the weights, runs inference, and outputs only the model response. Specifically, the code cannot output weights, the weight encryption key, or any information that is not directly outputted by the model.

Physical Security (ID.AM)

- **Increased guarding (compared with SL3) via manned and digital systems.** All access routes to the protected setup are guarded at all times with both manned and digital systems, so that no single individual or pair of individuals can physically access the protected setup without further approval or validation.
- **Meticulous logging of all access.** All access is both physically (visitor logbook) and digitally (security system logs) recorded, including who entered and when, without exceptions.
- **Prohibiting devices near the setup.** Digital devices (including laptops, phones, USB sticks, etc.) are not allowed inside or near the protected setup.

Permitted Interfaces (PR.AC)

- **Specialized hardware for all external interfaces.** All interfaces between the protected setup and the external world go through specialized hardware, which does not support running arbitrary code (e.g.,

[19] NVIDIA, *Confidential Compute on NVIDIA Hopper H100*, July 25, 2023.

field programmable gate arrays [FPGAs] or, for especially stable interfaces, application-specific integrated circuits [ASICs]), is very simple (not many features) and supports only a (simple) API for asking for inference (and, if necessary, for training/fine-tuning).

Monitoring (DE.CM)

- **Enforcement of time-buffered review (software limitation).** Require time limits on changes to the weight interface code (e.g., at least 24 hours from changelist proposal for review to approval and submission), such that multiple people are able to see the change and block it if it appears malicious. This helps prevent an attacker from using access to multiple employee devices to quickly (e.g., overnight) propose and approve malicious changes.
- **Protection of the monitoring logs at the hardware level.** Logs of the monitoring system (described in the previous level) are not editable at the hardware level (e.g., via unidirectional data diodes connecting to a separate [otherwise air-gapped] network).
- **Comprehensive anomaly detection and alert system over the monitoring logs.** Anomaly detection over these logs is comprehensive: All communications must be identifiable as one of a small set of well-understood behaviors and use cases. Any interaction outside the narrow bounds of previously seen and understood behavior raises an alert; when the interaction is understood, it is either added to the bounds of legitimate use or identified as a breach (whether malicious or unintentional) and responded to. This comprehensive approach can be applied to other parts of the system (including other logs, processes running on devices, and more); however, it may not be feasible to deploy for all of the organization's resources, thus necessitating prioritization.

AI Model Resilience

AI-specific security mitigations are generally nascent and fast-evolving. We expect recommendations in this space to change rapidly over the coming years. We include the items below to share the current state of the art, but we see these measures as less reliable than the majority of the measures mentioned.

Model Robustness (PR.DS)

- **Adversarial output detection.** A separate component (possibly an independent AI model) is used to classify outputs that seem to be the result of malicious activity and block them from being further processed or returned to the user.

Oracle Protection (PR.DS)

- **Output reconstruction.** After inference, outputs are randomly modified while minimizing effects on legitimate use—for example, by adding small noise to the logprobs or running outputs through an independent autoencoder.

Security of Network and Other (Nonweight) Sensitive Assets

Software (ID.AM)

- **Limiting the attack surface** (e.g., the limited interaction interfaces of a Chromebook). All devices with sensitive access have a vastly more limited attack surface than standard operating systems (e.g., Chromebook, QubesOS, or Bottlerocket for running containers, configured to optimize for security). In addition to the direct benefits of reducing the attack surface, this reduces noise during monitoring, making it easier for the security team to aggressively investigate any unexpected behavior.

Access, Permissions, and Credentials (PR.AC)

- **Enforcement of strong random passwords and keys for enhanced security.** Very strong passwords and keys (at least 128 bits of entropy, generated via strong random generation) are enforced.
- **Zero Trust architecture.** The organization adheres to at least the standards in the "Optimal" level of CISA's Zero Trust Maturity Model.[20]

Hardware (ID.AM)

- **All hardware used on devices must undergo source-code auditing and be validated as secure.** Network cards and GPUs are currently a particularly large gap.
- **Secure hardware required for access.** All devices are automatically verified before they are permitted access to the network.
- **Ongoing compromise assessment on all devices with access** (server or employee)—for example, via Mandiant, CrowdStrike, or FireEye.

Supply Chain (ID.SC)

- **Strict application allowlisting** (especially for sandboxes). Only trusted applications are allowed to execute on devices with access. Applications are not added without a thorough security review ensuring that the software does not have significant vulnerabilities, unless an important business need exists. The listing is continuously updated (adding new required applications and removing applications suspected of being insecure). Popular ML platforms likely would not pass a strict review because of their many dependencies; a very significant effort would be required to secure them or to build an in-house alternative.
- **SLSA Level 3 specification for all software used.**[21]
  - The SLSA specification does not include dependencies, so a certain artifact can live up to the SLSA guidelines while relying on non-SLSA compliant dependencies. The full dependency tree needs to be independently verified.
  - Note that SLSA (currently) only (noncomprehensively) ensures the provenance of the build process and does not ensure lack of tampering with source code or intentional malicious code by the provider. There is also no formal verification of SLSA. This should be viewed as a benefit to supply chain security but far from a guarantee.

Security Tooling (PR.PT)

- **Significant investment in advanced security systems.** Advanced security systems are put in place and significantly invested in, including
  - high-quality and heavily monitored honeypots
  - multiple security and monitoring systems that most employees have no way of seeing (e.g., they do not have access to the code and these systems are not documented anywhere visible to employees).

Physical Security (ID.AM)

- **Banning of unauthorized devices.** Phones and other non-authorized devices are not allowed near the network at different levels of strictness—e.g., guests are never allowed to bring in phones and no exceptions are allowed, employees cannot bring in phones but exceptions can be made. There are various technological options for monitoring such bans.

[20] CISA, 2023a.

[21] SLSA, homepage, undated.

Personnel Security

Filtering and Monitoring (PR.AT)

- **Preventing third-party access and reporting suspected illegitimate incidents.** There is a clear policy that third parties are never allowed access to weights or the protected setup, and there are clear instructions for all employees about how and to whom to report suspected illegitimate access (including outside the organization—e.g., to the FBI if a senior organization official is involved).
- **Advanced insider threat program.** The program has two tiers:
  - A program commensurate with the National Insider Threat Policy (or better) manages risk from any employees with access to internal organization resources (the network, codebase, etc.).[22] These are required by the U.S. government for the handling of classified data but are recommended to improve the security of CUI and other information in non–national security systems. The UK National Cyber Security Centre also provides recommendations for mitigating such risk.[23]
  - A program commensurate with the U.S. Personnel Reliability Program to ensure the reliability of those with direct access to the weights.[24] This would include some level of employee monitoring and, if possible, support from counterintelligence agencies.
- **Occasional employee integrity testing.** Although the predictive reliability of different integrity testing approaches is unclear, at least some forms of testing have been shown to deter employee misconduct.

Security Assurance and Testing

Red-Teaming and Penetration Testing (DE.AE)

- **Ongoing research and red-teaming to identify potential attack methods on the weight interface(s).**
- **Ensuring physical security through red-teaming.** In addition to the red-teaming of the digital security of the organization, the physical security of the weights' protected setup is reviewed and red-teamed to ensure top security standards.
- **Experience dealing with intelligence agencies.** In addition to significant experience with APTs, the third-party red-team has a proven track record of dealing with capable foreign intelligence services.

Risk and Security Assessments (DE.AE)

- **Automated weight exfiltration attempts.** Software is placed within the weights' protected setup that attempts to replicate itself and communicate with an external endpoint on the internet to continually test the weight network's isolation. In some circles, this is called *blue-dye testing.*
- **Manual weight exfiltration attempts.** A team would try to exfiltrate information by starting with the server or secure environment that the weights are stored in, searching for creative exfiltration paths, and addressing exfiltration routes identified.
- **Compliance with the FedRAMP High standards for security.** The organization implements controls commensurate with what is required by FedRAMP High (a U.S. federal government–wide compliance program that provides a standardized approach to security assessment, authorization, and continuous monitoring for cloud products and services).[25] This does not necessarily mean the organization must receive FedRAMP certification.

---

[22] White House, *National Insider Threat Policy*, 2011.

[23] National Cyber Security Centre, "Reducing Data Exfiltration by Malicious Insiders," June 30, 2022.

[24] Department of Defense Manual 5210.42, *Nuclear Weapons Personnel Reliability Program*, Department of Defense Under Secretary of Defense for Acquisition and Sustainment, January 13, 2015, incorporating change 4, May 9, 2022.

[25] FedRAMP, "Understanding Baselines and Impact Levels in FedRAMP," blog post, November 16, 2017.

Security Team Capacity (ID.RM)

- **General increased capacity (compared with SL3).** The security team has a capacity of about 100 people or more.
- **Greater concrete experience with APTs (compared with SL3).** The security team includes many (at least ten) people with independent real-world, hands-on experience with capable state actor-level APTs, including both offensive and defensive experience.
- **Zero-day vulnerability discovery capabilities.** The security team includes a significant number of top information security experts who can find novel globally influential zero-days.
- **The security team is empowered to not compromise security over other stakeholders.** The security team is given "unreasonable" power not to compromise on security to appease developers, product managers, leadership, users, etc.

Other Organization Policies (ID.RM)

- **Designating sensitive details of the weight security system.** Information about (at least significant components of) the weight security system is designated as sensitive, and policies are put in place to avoid leakage of such information—for example, securely storing relevant documents, and not discussing the weights security system outside locations considered secure.
- **Vetting of investors and other positions of influence.** Investors are thoroughly vetted to prevent inappropriate pressure undermining the security of the organization's assets. Although developed for a different purpose, the Defense Counterintelligence and Security Agency's (DCSA's) guidelines on Foreign Ownership, Control or Influence (FOCI) can help in identifying what roles and relationships would be vetted.[26]
- **Prioritizing leak prevention over other organizational goals.** Security becomes a top priority for the organization, to the extent that it sacrifices other goals to prevent leaks. The organization dedicates a nonnegligible portion of its annual budget to information security; the security team has veto (formally and in practice) over network, product, and work environment decisions that may undermine security (even if changes are important from a product or commercial perspective); and security is consistently prioritized over speed and flexibility across organization decisions.
- **Four independent security layers.** The security team maintains four independent security layers, as defined in SL3. The larger number of independent security layers is critical for making the system robust against actors expected to have many zero-days and other advantages relative to defenders.

[26] Defense Counterintelligence and Security Agency, "Foreign Ownership, Control or Influence," webpage, undated.

Security Level 5 (SL5)

Implementation of Previous Security Levels

- The organization has implemented all the controls from SL1–SL4.

Weight Security

Weight Storage (PR.DS)

- **Extreme isolation of weight storage.** All copies of the weights are stored in an isolated environment that adheres to one of the following two requirements:
  - **Storage alternative A: Completely isolated network.** Any extraction of information from the completely isolated network (including USB devices, hard drives, physical documents) must be stringently reviewed and approved by multiple people. It is a rare occurrence. Information is encrypted using well-known and commonly used (at least 256-bit encryption) easy-to-destroy devices that have limited capacity for information storage (at least two orders of magnitude smaller than the capacity to store the full weights). Optical media may be a good candidate for this. Regardless of the process, no more than 50GB of data can leave the completely isolated network per month in aggregate across all forms of communication. An important exception is extraction of the weights themselves through an HSM-like device (see the next bullet). We expect the training environment and part or all of the research environment (at least those parts that require direct access to the weights, such as some types of interpretability research) to be hosted in a completely isolated network environment. This may change based on need.
  - **Storage alternative B: HSM variants for ML.** Weights are stored in a protected environment as required in SL4. Within that protected environment, they are stored (only) in specialized HSM variants developed for ML applications.
    - The weights themselves (and not just their decryption keys) are stored only within the TEE.
    - The TEE supports fine-tuning and inference, but nothing more, and, in particular, cannot output the weights. The fine-tuning interface must be limited: Loss function and other configurations must be limited so that the weights cannot be extracted via the amount of fine-tuning and inference requests expected throughout the normal lifetime of such a device. These constraints are enforced at the hardware level, not only through the validation of signed code.
    - The TEE is enclosed by a state-of-the-art tamperproof enclosure to ensure it is secure against advanced physical attacks. The enclosure conforms to FIPS 140-3 Security Level 6.[27] Conforming to this standard may not be a sufficient requirement, so we recommend receiving specialized guidance from the relevant national security agencies.
    - Drawing on historical cases of extremely sensitive digital information that is needed to perform services at scale (e.g., private keys of root certificate authorities), we suggest that this approach is far more robust than other approaches we describe in this report: It reduces the attack surface much more than other solutions, no individual or group of individuals can access the weights in any way other than limited and predetermined interactions, it provides better protection against physical attacks, and it seems to be less vulnerable to configuration errors and non-implementation or mis-implementation of policy. However, it does not protect against abuse of legitimate APIs, which is addressed under the "Permitted Interfaces" and "AI Model Resilience" sections that follow.

[27] NIST, *Security Requirements for Cryptographic Modules*, Federal Information Processing Standards Publication 140-3, March 22, 2019.

        - We expect all production deployments (public API, internal, and on-premises) to use **specialized HSMs** (this may change based on need). Production purposes would require many copies of such devices to support many instances of the model. Such devices may need to communicate with each other efficiently and be deeply integrated with the training and inference infrastructure to be effective in addition to being secure.
        - We are unaware of an existing implementation of such an approach, and many implementation questions remain. Many of these questions involve trade-offs between usability and security; for example, how much flexibility (in internal network architecture, loss function, etc.) should be enabled, whether and how to allow copying the weights to other device copies (e.g., to allow quick and flexible scale-up for production purposes), how and whether syncing of weights across GPUs should be supported during training, how extraction of weights through gradients and inference results can be prevented, and more.
    - Note that if the initial training of the model is infeasible to do within the HSM (and is therefore performed in a completely isolated network), the requirement for weights to be in one of these environments at all times imply that they must be inserted into the HSM-variant within the completely isolated network and the HSM-variant is the only storage device that can leave the isolated network.
- **Advanced preventive measures for side-channel attacks.** External connections to the protected environment (in the non-isolated setup), which are done through dedicated hardware as described for SL4, also inject noise, time delays, and other tools to prevent side-channel attacks, exfiltration, and creative state-of-the-art attacks.
- **Formal hardware verification of key components.** Hardware components that are critical for the security of the weights (e.g., all the above, the interface between the protected setup described in SL4, or other alternatives adapted to the organization) have their correctness formally verified.

Physical Security (ID.AM)

- **Increased significant guarding** (compared with SL4). Protected areas are secured by multiple armed guards and digital security systems at all times.
- **Supervised access for everyone.** No one (including authorized personnel) can interact with the system alone without other trusted authorized personnel.
- **Routine rigorous device inspections.** All devices within the completely isolated network are inspected meticulously (for drill holes, unauthorized devices, bugs, replaced cables, etc.) and frequently (e.g., weekly).
- **Disabling of most communication at the hardware level.** Devices in the completely isolated network that are hardened as recommended in SL4 (camera, speakers, Wi-Fi module, antennae, USB ports, etc.) are all physically removed or disabled, rather than disabled by configuration or software.

Permitted Interfaces (PR.AC)

- **Strict limitation of external connections to the completely isolated network.** External connections to the completely isolated network are approved only for production-critical uses. All other interactions with the system are done locally and in person to reduce the digital attack surface.

Access Control (PR.AC)

- **Irrecoverable key policy.** There are no alternative access or key retrieval systems for all keys relevant to encryption or access to weights. If the password or security key is lost, those data are lost forever, and there is no alternative system to escalate a request for access.

Standard Compliance (ID.GV)

- **Protection equivalent to that required for TS/SCI.** Any system that contains or has access to the weights has equivalent protection to that required for TS/SCI. Some relevant specifications are available in the National Counterintelligence and Security Center's ICD/ICS 705: *Technical Specifications for Construction and Management of SCIFs,* but relevant government bodies should be consulted for the full specifications.[28] Accreditation by the government is not necessary.

AI Model Resilience

AI-specific security mitigations are generally nascent and fast-evolving. We expect recommendations in this space to change rapidly over the coming years. We include the items below to share the current state of the art, but we see these measures as less reliable than the majority of the measures mentioned.

Oracle Protection (PR.DS)

- **Constant inference time.** To protect against timing attacks, computation running time does not depend on the weights' values.

Security of Network and Other (Nonweight) Sensitive Assets

Supply Chain (ID.SC)

- **Strong limitations on software providers.** All software is built internally or developed and/or recommended by an extremely reliable and discriminating source (e.g., developed by the national security community).
- **Strong limitations on hardware providers.** Any hardware that is not developed internally must be thoroughly and continuously vetted and recommended by highly reliable, well-informed, and discriminating sources (e.g., the national security community provides a short list of highly trusted providers).

Personnel Security

Personal Protection (ID.AM)

- **Proactive protection of executives and individuals handling sensitive materials.** All senior officials in the organization and individuals with access to sensitive materials receive physical protection (e.g., security guards), counterintelligence support, and other means of ensuring that they are safe and are less likely to be threatened or extorted.

Security Assurance and Testing

Red-Teaming and Penetration Testing (DE.AE)

- **Proactive search for crucial vulnerabilities.** There is a dedicated team of top talent searching for new zero-days and other vulnerabilities that might be relevant to the system (like Google's Project Zero but with an emphasis on protecting weights).

---

[28] National Counterintelligence and Security Center, *Technical Specifications for Construction and Management of Sensitive Compartmented Information Facilities,* version 1.5.1, July 26, 2021.

Maintenance (PR.MA)

- **Security is strongly prioritized over availability.** Security of the weights is defined to be a significantly higher priority than system availability: Security policies are never breached, even if the system is down (e.g., connecting external devices to the completely isolated network to debug a critical production issue).

Other Organization Policies (ID.RM)

- **Eight independent security layers.** The security team maintains eight independent security layers. The larger number of independent security layers is critical for making the system robust against actors that can produce zero-days and unexpected solutions at a high pace.

APPENDIX C

Summary of Security Measures Across Levels

Table C.1 summarizes all security measures across all security levels and all categories.

TABLE C.1
Summary of Security Measures Across Levels

Category	Subcategory	SL1	SL2	SL3	SL4	SL5
Implementation of Previous Security Levels	–	–	• The organization has implemented all controls from SL1.	• The organization has implemented all controls from SL1 and SL2.	• The organization has implemented all controls from SL1–SL3.	• The organization has implemented all controls from SL1–SL4.
Weight Security	Weight Storage	• Sensitive data remain internal. • Weight encryption (best effort)	• Storage location (e.g., weights are stored exclusively on servers and not on local devices) • Encryption (e.g., all keys are secured in a key management system)	• Centralized and restricted management of weight storage • Secure cloud network (if applicable) • Dedicated devices for weights and weight security data	• Isolation of weight storage • Weight storage setup is protected against eavesdropping and the simplest of TEMPEST attacks. • Hardware-enforced limits on output rate • Reduce communication capabilities	• Extreme isolation of weight storage (completely isolated network) • Advanced preventive measures for side-channel attacks (e.g., noise injection) • Formal hardware verification of key components
Weight Security	Security During Transport and Use	–	• Encryption in transit (e.g., not transporting weights over public or unencrypted channels)	–	• Confidential computing (when available)	–
Weight Security	Physical Security	• Data centers of cloud providers	• Data centers are guarded, and only people with authorization are allowed inside. • Visitor access is restricted and logged.	• Data centers are guarded or locked at all times. • Premises are swept for intruders frequently. • Premises are meticulously swept for unauthorized devices routinely.	• Increased guarding (compared with SL3) via manned and digital systems • Meticulous logging of all access • Prohibiting devices near the setup	• Increased significant guarding (compared with SL4) • Supervised access for everyone • Routine rigorous device inspections • Disabling of most communication at the hardware level

Table C.1—Continued

Category	Subcategory	SL1	SL2	SL3	SL4	SL5
Weight Security	Permitted Interfaces	–	–	• Authorized users who interact with the weights do so through a software interface that reduces risk of the weights being illegitimately copied. • Any code accessing the weights minimizes attack surface, provides only simple forms of access, and uses the minimal amount of (highly trusted and well-established) external code necessary. • Avoiding model interactions that bypass monitoring or constraints	• Specialized hardware for all external interfaces	• Strict limitation of external connections to the completely isolated network
Weight Security	Access Control	• Access control for sensitive assets • Access log or audit trail	• Restrictions on sensitive interactions (e.g., require multifactor authentication using FIDO authentication/ hardware security keys)	• Protocols and policies for sensitive interactions (e.g., access to the various permitted interfaces to the weights is stringently controlled, multiparty authorization, security reviews, etc.)	–	• Irrecoverable key policy (barring alternative access or key retrieval systems)
Weight Security	Monitoring	–	• Logging of all sensitive interactions • Regulation and monitoring of weight copies across the organization network	• Ongoing manual monitoring of sensitive interactions • Ongoing automated anomaly detection • Automated and manual monitoring/ blocking of potentially malicious queries • Frequent compromise assessment • Frequent integrity checks via comparison against a baseline system configuration ("gold image")	• Enforcement of time-buffered review (software limitation) • Protection of the monitoring logs at the hardware level • Comprehensive anomaly detection and alert system over the monitoring logs	–
Weight Security	Standard Compliance	–	–	• Implementation of measures described in NIST SP 800-171 or equivalent • Future implementation of measures described by CMMC 2.0 Level 3	• Compliance with standard protocols for sensitive governmental information	• Protection equivalent to that required for TS/ SCI

Table C.1—Continued

Category	Subcategory	SL1	SL2	SL3	SL4	SL5
AI Model Resilience	Model Robustness	–	• Input reconstruction (e.g., during inference, a privately known prefix is added ahead of the user prompt) • Adversarial training	• Adversarial input detection	• Adversarial output detection	–
AI Model Resilience	Oracle Protection	–	–	• Limitations on the number of inferences using the same credentials	• Output reconstruction	• Constant inference time
Security of Network and Other (Nonweight) Sensitive Assets	Software	• Moderately frequent software update management and compliance monitoring	• Frequent software update management and compliance monitoring	• Very frequent software update management and compliance monitoring	• Limiting the attack surface (e.g., the limited interaction interfaces of a Chromebook)	–
Security of Network and Other (Nonweight) Sensitive Assets	Access, Permissions, and Credentials	• Least privilege principle • Restrictions on device and account sharing • Password best practices • Multifactor authentication • Single Sign-On (SSO) • Backup and recovery tools • Commercial identity and access management (IAM) tools • Zero Trust architecture (adherence to at least the standards in the "Traditional" level of CISA's Zero Trust Maturity Model)	• Strong password enforcement • The work network is separate from the guest network. • Guest accounts disabled whenever possible • Strong access management tools • Zero Trust architecture (adherence to at least the standards in the "Initial" level of CISA's Zero Trust Maturity Model)	• 802.1x authentication • Zero Trust architecture (adherence to at least the standards in the "Advanced" level of CISA's Zero Trust Maturity Model)	• Enforcement of strong random passwords and keys for enhanced security • Zero Trust architecture (adherence to at least the standards in the "Optimal" level of CISA's Zero Trust Maturity Model)	–
Security of Network and Other (Nonweight) Sensitive Assets	Hardware	• Modern device architectures that establish root of trust and block malicious code execution • CPU anti-exploitation features	• Lost or stolen devices reported • All network devices are visible and trackable.	• Security-minded hardware sourcing	• All hardware used on devices must undergo source-code auditing and be validated as secure. • Secure hardware required for access • Ongoing compromise assessment on all devices with access	–

Table C.1—Continued

Category	Subcategory	SL1	SL2	SL3	SL4	SL5
Security of Network and Other (Nonweight) Sensitive Assets	Supply Chain	• The reputability of software is reviewed before incorporation.	• Review of vendor and supplier security	• Software inventory management • Supply chain security is commensurate with the organization's security	• Strict application allowlisting (especially for sandboxes) • SLSA Level 3 specification for all software used	• Strong limitations on software providers (e.g., built internally or developed and/or recommended by an extremely reliable and discriminating source) • Strong limitations on hardware providers (e.g., hardware that is not developed internally must be thoroughly and continuously vetted)
Security of Network and Other (Nonweight) Sensitive Assets	Security Tooling	• Modern authentication infrastructure • Commercial network security solutions • Commercial endpoint security solutions • Reliance on standard security infrastructure (depending on circumstances)	• Disk encryption • Network communications are encrypted by default. • Email security tools • Use of integrated security approaches	• Enforcement of security policies through code rather than manual compliance • Security policy enforcement for network access across devices	• Significant investment in advanced security systems	–
Security of Network and Other (Nonweight) Sensitive Assets	Configuration Management	• Enforce screen locks for inactivity	• Incorporate fundamental infrastructure and policies for Security-by-Design and Security-by-Default • Configuration management monitoring	–	–	–
Security of Network and Other (Nonweight) Sensitive Assets	Physical Security	–	• Office security • Careful disposal of printed materials	–	• Banning of unauthorized devices	–
Personnel Security	Awareness and Training	• Basic onboarding information security training for employees	• Periodic mandatory information security training for all employees • Employee training on configuration errors and their security implications	• Employee awareness of weight interaction monitoring • Security training for employees (not necessarily only those with access) • Security risk reporting program	–	–
Personnel Security	Filtering and Monitoring	–	• Installation of monitoring software for secure network access • Active drills to identify and educate noncompliant employees	• Insider threat program	• Preventing third-party access and reporting suspected illegitimate incidents • Advanced insider threat program • Occasional employee integrity testing	–

Table C.1—Continued

Category	Subcategory	SL1	SL2	SL3	SL4	SL5
Personnel Security	Personal Protection	–	–	–	–	• Proactive protection of executives and individuals handling sensitive materials
Security Assurance and Testing	Red-Teaming and Penetration Testing	–	• Mandatory external reviews	• Ongoing penetration testing • Penetration testing of physical access and facility security • Advanced red-teaming: – Elite external team – Substantial funding – Access to design and code – Testing insider threats – Expanded access – Attention to the weights and authentication	• Ongoing research and red-teaming to identify potential attack methods on weight interface(s) • Ensuring physical security through red-teaming • Experience dealing with intelligence agencies	• Proactive search for crucial vulnerabilities (e.g., zero-days)
Security Assurance and Testing	Risk and Security Assessments	• Internal reviews	–	• Keeping a risk register	• Automated weight exfiltration attempts • Manual weight exfiltration attempts • Compliance with the FedRAMP High standards for security	–
Security Assurance and Testing	Community Involvement and Reporting	–	• Bug-bounty and vulnerability-discovery programs	–	–	–
Security Assurance and Testing	Software Development Process	–	• Secure software development standards (compliance with NIST's Secure Software Development Framework)	–	–	–
Threat Detection and Response	–	–	• Protocols and funding for rapid incident response • Incident reporting	• Placement of effective honeypots	–	–
Security Team Capacity	–	• Basic incident response capabilities	• Constant availability of qualified personnel	• General increased capacity (compared with SL2) • Concrete experience with APTs • Leveraging diverse security experience from leading organizations	• General increased capacity (compared with SL3) • Greater concrete experience with APTs (compared with SL3) • Zero-day vulnerability discovery capabilities • The security team is empowered not to compromise security over other stakeholders.	–

Table C.1—Continued

Category	Subcategory	SL1	SL2	SL3	SL4	SL5
Maintenance	–	• Information security news monitoring and implementation	• Continuous vulnerability management and adaptation to information security developments	–	–	• Security is strongly prioritized over availability (e.g., barring connecting external devices to the completely isolated network to debug a critical production issue)
Other Organization Policies	–	–	• Promotion of a security mindset by organization management • Stringent remote work policies	• Two independent security layers	• Designating sensitive details of the weights security system • Vetting of investors and other positions of influence • Prioritizing leak prevention over other organizational goals • Four independent security layers	• Eight independent security layers

APPENDIX D

Informal Interview Protocol

This appendix lists the questions that were used in our initial interviews eliciting relevant attack vectors, security measures, and priorities for the AI organizations. This is not a formal or full protocol; the phrasing varied depending on context, full discussion followed up on answers to these questions, and sometimes not all questions were asked (questions were prioritized based on the expertise of the specific expert).

Following a short description of the goals of the report, the following questions were asked:

- Any thoughts about them?
- The report's scope?
- Other strategic considerations in how we should be thinking about this?
- Any attack surfaces, threat vectors, or approaches threat actors can take that organizations developing frontier models should be addressing?
  - Any that are unique to the protection of AI model weights, or are particularly relevant to specific environments AI models are used in (e.g., research, training, production, internal deployment, on-premises deployment . . .)
  - Any that you believe are important and underappreciated or not sufficiently secured against or considered in AI organizations' threat models?
  - Any (publicly available) examples of specific attack vectors being successfully used in the real world (especially more controversial or advanced ones)?
- Any concrete security measures you believe should be a high priority for AI organizations to implement to secure the weights of frontier AI models?
  - Any that are unique to the protection of AI model weights, or are particularly relevant to specific environments in which AI models are used in (e.g., research, training, production, internal deployment, on-premises deployment . . .)
  - Any that become relevant only when the threat model expands to actors with additional capabilities (e.g., APTs, insider threats, state actors, etc.)?
  - Any broader principles, approaches, or advice, even if they are not a specific and concrete security measure?
  - What would you most advocate for, taking into account both the security benefits and the costs (including implementation, productivity harms, etc.)?
- Any thoughts on some of the prominent or more controversial attack vectors or security measures we've already heard? [once we had such prior suggestions]
- Other existing discussions around securing AI model weights that we should be aware of? Areas of consensus or points of contention?
- What would make the report more useful for you?
- Is there anything else you would like to share, or you think we should include in the report?
- Who else should we talk to in order to inform the report?
- Recommendations for resources we should be reviewing for the report?

- Would you be open to reviewing drafts of our analysis and providing feedback?

Follow-up engagements did not follow this protocol but instead either dived into topics brought up in the initial interview, consulted on specific suggestions or estimates, or shared sections of our analysis and elicited feedback, recommendations, and corrections. This included receiving copies of the operational capacity categories to review and suggest changes; the attack vectors to add or contest specific attack vectors; the capability assessment table to fill in or suggest changes to the feasibility scores; and different iterations of the security level benchmarks to suggest additions, removals, changes to the assortment into levels, or definitions and requirements of each of the security measures.

Abbreviations

AI	artificial intelligence
API	application program interface
APT	advanced persistent threat
ATLAS	Adversarial Threat Landscape for Artificial-Intelligence Systems
AWS	Amazon Web Services
BIML	Berryville Institute of Machine Learning
CA	certification authority
CIA	Central Intelligence Agency
CISA	Cybersecurity and Infrastructure Security Agency
CMMC	Cybersecurity Maturity Model Certification
CPU	central processing unit
CUI	Controlled Unclassified Information
CVE	common vulnerability and exposure
FBI	Federal Bureau of Investigation
FIDO	Fast Identity Online
GPT	generative pre-trained transformer
GPU	graphics processing unit
HSM	hardware security module
HTTP	Hypertext Transfer Protocol
HVAC	heating, ventilation, and air conditioning
IAM	identity and access management
IMAP	Internet Message Access Protocol
IT	information technology
LLM	large language model
ML	machine learning
NASA	National Aeronautics and Space Administration
NIST	National Institute of Standards and Technology
NSA	National Security Agency
OC	operational capacity
OPM	Office of Personnel Management
PC	personal computer
PII	personally identifiable information
R&D	research and development
RAM	random-access memory
RSA	Rivest-Shamir-Adleman
SCI	Sensitive Compartmented Information
SIM	subscriber identity module
SL	security level

SLSA	Supply-chain Levels for Software Architects
SMB	server message block
SSH	secure remote shell
SSL	Secure Sockets Layer
TCP/IP	Transmission Control Protocol/Internet Protocol
TEE	trusted execution environment
TLS	Transport Layer Security
TS	Top Secret
UEFI	Unified Extensible Firmware Interface
UN	United Nations
USB	universal serial bus
VPN	virtual private network
WEP	Wired Equivalent Privacy

References

"2 Navy Sailors Arrested, Accused of Providing China with Information," CBS News, updated August 3, 2023.

Aarness, Anne, "What Is XDR? How to Choose an XDR Solution," *CrowdStrike*, April 18, 2023.

Ablon, Lillian, and Andy Bogart, *Zero Days, Thousands of Nights: The Life and Times of Zero-Day Vulnerabilities and Their Exploits*, RAND Corporation, RR-1751-RC, 2017. As of February 15, 2024: https://www.rand.org/pubs/research_reports/RR1751.html

Abrams, Lawrence, "Fortinet Fixes Critical RCE Flaw in Fortigate SSL-VPN Devices, Patch Now," *Bleeping Computer*, June 11, 2023.

Adkins, Heather, Betsy Beyer, Paul Blankinship, Piotr Lewandowski, Ana Oprea, and Adam Stubblefield, *Building Secure and Reliable Systems: Best Practices for Designing, Implementing, and Maintaining Systems*, O'Reilly, 2020.

Adrian, David, Karthikeyan Bhargavan, Zakir Durumeric, Pierrick Gaudry, Matthew Green, J. Alex Halderman, Nadia Heninger, Drew Springall, Emmanuel Thomé, Luke Valenta, Benjamin VanderSloot, Eric Wustrow, Santiago Zanella-Béguelin, and Paul Zimmermann, *Imperfect Forward Secrecy: How Diffie-Hellman Fails in Practice*, 22nd ACM Conference on Computer and Communications Security (CCS '15), October 2015.

Afifi-Sabet, Keumars, "NASA Employee Data Exposed for at Least Three Weeks Due to Misconfigured Web App," *IT Pro*, January 14, 2019.

Anderson, Ross, *Security Engineering: A Guide to Building Dependable Distributed Systems*, 3rd ed., Wiley, 2020.

Andrews, Wilson, and Todd Lindeman, "$52.6 Billion: The Black Budget," *Washington Post*, August 29, 2013.

Anthropic, "Anthropic's Responsible Scaling Policy," company announcement, September 19, 2023.

Arghire, Ionut, "Millions of Devices Remain Exposed via SMB, Telnet Ports: Rapid7," *Security Week*, June 15, 2017.

Armis Security, "Two Years in and WannaCry Is Still Unmanageable," May 29, 2019. As of February 15, 2024: https://www.armis.com/blog/two-years-in-and-wannacry-is-still-unmanageable/

Artificial Intelligence Index, *Artificial Intelligence Index Report 2024*, Stanford University, 2024.

AttackerKB, CVE year 2022 search results sorted by value, webpage, undated-a. As of February 15, 2024: https://attackerkb.com/search?attackerValue=4%2C5&tags=exploitedInTheWild&cveYear=2022

AttackerKB, CVE year 2022 search results, webpage, undated-b. As of February 15, 2024: https://attackerkb.com/search?attackerValue=&tags=exploitedInTheWild&cveYear=2022

AttackerKB, homepage, undated-c. As of February 15, 2024: https://attackerkb.com/

Bals, Fred, "CVE-2017-5638: The Apache Struts Vulnerability Explained," Synopsys, blog post, September 13, 2017. As of February 20, 2024: https://www.synopsys.com/blogs/software-security/cve-2017-5638-apache-struts-vulnerability-explained.html

Barnes, Joe, "Russia Forces Ukrainians to Become Spies by Threatening to Kill Their Families," *The Telegraph*, February 6, 2024.

Barracuda MSP, *Spear Phishing: Top Threats and Trends*, Vol. 7, March 2022.

Beer, Ian, and Samuel Groß, "A Deep Dive into an NSO Zero-Click iMessage Exploit: Remote Code Execution," Google Project Zero blog post, December 15, 2021. As of February 15, 2024: https://googleprojectzero.blogspot.com/2021/12/a-deep-dive-into-nso-zero-click.html

Beierle, Christof, Patrick Derbez, Gregor Leander, Gaëtan Leurent, Håvard Raddum, Yann Rotella, David Rupprecht, and Lukas Stennes, "Cryptanalysis of the GPRS Encryption Algorithms GEA-1 and GEA-2," Paper 2021/819, *Eurocrypt 2021*, International Association for Cryptologic Research, 2021.

Belanger, Ashley, "SIM-Swapping Ring Stole $400M in Crypto from a US Company, Officials Allege," *Ars Technica*, January 30, 2024.

Berger, Andreas, "What Is Log4Shell? The Log4j Vulnerability Explained (and What to Do About It)," Dynatrace, blog post, updated June 1, 2023. As of February 15, 2024: https://www.dynatrace.com/news/blog/what-is-log4shell/

Bernstein, Daniel J., and Tanja Lange, "SafeCurves: Choosing Safe Curves for Elliptic-Curve Cryptography," webpage, December 1, 2014. As of April 2, 2024: https://cr.yp.to/talks/2014.01.18/slides-dan+tanja-20140118-a4.pdf

Berryville Institute of Machine Learning, "BIML Interactive Machine Learning Risk Framework," webpage, undated. As of February 15, 2024: https://berryvilleiml.com/interactive/

Bıçakcı, Salih, *Introduction to Cyber Security for Nuclear Facilities*, Centre for Economics and Foreign Policy Studies, 2015.

BIML—*See* Berryville Institute of Machine Learning.

"BLASTPASS: NSO Group iPhone Zero-Click, Zero-Day Exploit Captured in the Wild," Citizen Lab, Munk School of Global Affairs & Public Policy, University of Toronto, September 7, 2023.

Borisov, Nikita, Ian Goldberg, and David Wagner, "Intercepting Mobile Communications: The Insecurity of 802.11," *MobiCom '01: Proceedings of the 7th Annual International Conference on Mobile Computing and Networking*, July 2001.

Boyle, Gage, and Kenny Paterson, *20 Years of Bleichenbacher's Attack*, Royal Holloway University of London, ISG MSc Information Security thesis series, 2019.

Brodskey, Zev, "The Capital One Data Breach: How Crisis Could Have Been Averted," Perimeter 81, July 31, 2019.

Brumley, Billy, *A3/A8 & COMP128*, T-79.514 Special Course on Cryptology, Helsinki University of Technology, November 11. 2004.

Calyptix Security, "DNC Hacks: How Spear Phishing Emails Were Used," blog post, December 30, 2016. As of February 20, 2024: https://www.calyptix.com/news-and-events/dnc-hacks-how-spear-phishing-emails-were-used/

Carlini, Nicholas, Daniel Paleka, Krishnamurthy Dvijotham, Thomas Steinke, Jonathan Hayase, A. Feder Cooper, Katherine Lee, Matthew Jagielski, Milad Nasr, Arthur Conmy, et al., "Stealing Part of a Production Language Model," arXiv, arXiv:2403.06634, March 11, 2024.

Chief Information Officer, U.S. Department of Defense, "CMCC Model," webpage, undated. As of February 20, 2024: https://dodcio.defense.gov/CMMC/Model/

Chilson, Neil, "A Lesson from Uber: Secure Your Non-Production Software Environments," Federal Trade Commission, blog post, April 12, 2018. As of February 20, 2024: https://www.ftc.gov/policy/advocacy-research/tech-at-ftc/2018/04/lesson-uber-secure-your-non-production-software-environments

Chui, Michael, Eric Hazan, Roger Roberts, Alex Singla, Kate Smaje, Alex Sukharevsky, Lareina Yee, and Rodney Zemmel, *The Economic Potential of Generative AI: The Next Productivity Frontier*, McKinsey & Company, June 14, 2023.

Cichonski, Paul, Thomas Millar, Tim Grance, and Karen Scarfone, *Computer Security Incident Handling Guide*, NIST SP 800-61 Rev. 2, National Institute of Standards and Technology, August 2012. As of February 15, 2024: https://csrc.nist.gov/pubs/sp/800/61/r2/final

Cimpanu, Catalin, "Hacker Ransoms 23k MongoDB Databases and Threatens to Contact GDPR Authorities," *ZDNet*, July 1, 2020.

Cimpanu, Catalin, "Cobalt Strike and Metasploit Accounted for a Quarter of All Malware C&C Servers in 2020," *ZDNet*, January 7, 2021a.

Cimpanu, Catalin, "First Fully Weaponized Spectre Exploit Discovered Online," *The Record*, March 10, 2021b.

CISA—*See* Cybersecurity and Infrastructure Security Agency.

Clark, Mitchell, Richard Lawler, and Jay Peters, "Microsoft Confirms Lapsus$ Hackers Stole Source Code via 'Limited' Access," *The Verge*, March 22, 2022.

Cloudflare, "What Is Defense in Depth? Layered Security," webpage, undated-a. As of April 3, 2024: https://www.cloudflare.com/learning/security/glossary/what-is-defense-in-depth/

Cloudflare, "What Was the WannaCry Ransomware Attack?" webpage, undated-b. As of February 15, 2024: https://www.cloudflare.com/learning/security/ransomware/wannacry-ransomware/

"Collide+Power, Downfall, and Inception: New Side-Channel Attacks Affecting Modern CPUs," *Hacker News*, August 9, 2023.

Comodo Cybersecurity, "Update 31-MAR-2011," webpage, March 31, 2011. As of May 10, 2024: https://www.comodo.com/Comodo-Fraud-Incident-2011-03-23.html

Computer Emergency Response Team, National Insider Threat Center, *Common Sense Guide to Mitigating Insider Threats*, Carnegie Mellon University, 2022.

Cooney, Michael, "Cisco Talos Details Exceptionally Dangerous DNS Hijacking Attack," *Network World*, April 17, 2019.

Council on Foreign Relations, "Operation Aurora," webpage, undated. As of February 15, 2024: https://www.cfr.org/cyber-operations/operation-aurora

Cox, Joseph, "Microsoft Employees Exposed Own Company's Internal Logins," *Vice*, August 16, 2022.

"Cybercriminals Increasingly Using EvilProxy Phishing Kit to Target Executives," *Hacker News*, August 10, 2023.

Cybersecurity and Infrastructure Security Agency, "Stuxnet Malware Mitigation (Update B)," Alert Code ICSA-10-238-01B, updated January 8, 2014.

Cybersecurity and Infrastructure Security Agency, "Targeted Destructive Malware," Alert Code AA21-008A, updated January 3, 2020a.

Cybersecurity and Infrastructure Security Agency, *Insider Threat Mitigation Guide*, November 2020b.

Cybersecurity and Infrastructure Security Agency, "Detecting Post-Compromise Threat Activity in Microsoft Cloud Environments," Alert Code TA14-353A, updated April 15, 2021a.

Cybersecurity and Infrastructure Security Agency, "Mitigating Log4Shell and Other Log4j-Related Vulnerabilities," Advisory, Alert Code AA21-356A, updated December 23, 2021b.

Cybersecurity and Infrastructure Security Agency, "Implementing Phishing-Resistant MFA," fact sheet, October 2022.

Cybersecurity and Infrastructure Security Agency, *Zero Trust Maturity Model*, version 2.0, April 2023a.

Cybersecurity and Infrastructure Security Agency, *Secure-by-Design: Shifting the Balance of Cybersecurity Risk: Principles and Approaches for Secure by Design Software*, October 25, 2023b.

"Data on 123 Million US Households Exposed Due to Misconfigured AWS S3 Bucket," *Trend Micro*, December 20, 2017.

Debusmann, Bernd, Jr., "Typo Sends Millions of US Military Emails to Russian Ally Mali," BBC News, July 17, 2023.

Defense Counterintelligence and Security Agency, "Foreign Ownership, Control or Influence," webpage, undated. As of February 21, 2024:
https://www.dcsa.mil/Industrial-Security/Entity-Vetting-Facility-Clearances-FOCI/Foreign-Ownership-Control-or-Influence/

Defense Security Service and National Counterintelligence and Security Center, *Counterintelligence: Best Practices for Cleared Industry*, undated.

Department of Defense Manual 5210.42, *Nuclear Weapons Personnel Reliability Program*, Under Secretary of Defense for Acquisition and Sustainment, January 13, 2015, incorporating change 4, May 9, 2022.

"Did Chinese Hack Cabinet Secretary's Laptop?" NBC News, May 29, 2008.

Dobson, Melina, Jason Dymydiuk, and Sarah Mainwaring, "Operation Rubicon: The Most Successful Intelligence Heist of the 20th Century," Warwick Knowledge Center, undated.

Dorais-Joncas, Alexis, and Facundo Muñoz, *Jumping the Air Gap: 15 Years of Nation-State Effort*, ESET, December 2021a.

Dowd, Mark, "How Do You Actually Find Bugs?" keynote video, April 21, 2022. As of February 15, 2024: https://m.youtube.com/watch?v=7Ysy6iA2sqA

Dutta, Sankha Baran, Hoda Naghibijouybari, Arjun Gupta, Nael Abu-Ghazaleh, Andres Marquez, and Kevin Barker, "Spy in the GPU-Box: Covert and Side Channel Attacks on Multi-GPU Systems," *Proceedings of the 50th Annual International Symposium on Computer Architecture*, 2023.

Easter, David, "The Impact of 'Tempest' on Anglo-American Communications Security and Intelligence, 1943–1970," *Intelligence and National Security*, Vol. 36, No. 1, 2021.

Eddy, Nathan, "Cloud Misconfig Exposes 3TB of Sensitive Airport Data in Amazon S3 Bucket: 'Lives at Stake,'" *Dark Reading*, July 6, 2022.

Electronic Privacy Information Center, "Vulnerabilities Equities Process," webpage, undated. As of February 15, 2024:
https://archive.epic.org/privacy/cybersecurity/vep/

Epoch AI, "Announcing Epoch AI's Updated Parameter, Compute and Data Trends Database," October 23, 2023. As of May 8, 2024:
https://epochai.org/blog/announcing-updated-pcd-database

European Parliament, "General Data Protection Regulation," webpage, 2016. As of February 15, 2024: https://gdpr-info.eu/

Executive Order 14091, "Further Advancing Racial Equity and Support for Underserved Communities Through the Federal Government," Executive Office of the President, February 16, 2023.

Fadilpašić, Sead, "PyTorch Hit by Severe Security Compromise," *TechRadar*, January 3, 2023.

Faife, Corin, "Uber's Hack Shows the Stubborn Power of Social Engineering," *The Verge*, September 16, 2022.

FBI—*See* Federal Bureau of Investigation.

Federal Bureau of Investigation, "Ana Montes: Cuban Spy," webpage, undated-a. As of May 1, 2024: https://www.fbi.gov/history/famous-cases/ana-montes-cuba-spy

Federal Bureau of Investigation, "Robert Hanssen," webpage, undated-b. As of May 2, 2024: https://www.fbi.gov/history/famous-cases/robert-hanssen

Federal Bureau of Investigation, Chicago Division, "Suburban Chicago Woman Sentenced to Four Years in Prison for Stealing Motorola Trade Secrets Before Boarding Plane to China," August 29, 2012.

FedRAMP, "Understanding Baselines and Impact Levels in FedRAMP," blog post, November 16, 2017. As of April 30, 2024:
https://www.fedramp.gov/understanding-baselines-and-impact-levels/

Files, John, "V.A. Laptop Is Recovered, Its Data Intact," *New York Times*, June 30, 2006.

Fingas, Jon, "Stuxnet Worm Entered Iran's Nuclear Facilities Through Hacked Suppliers," *Engadget*, November 13, 2014.

Forbes, Tom, "I Scanned Every Package on PyPi and Found 57 Live AWS Keys," blog post, January 6, 2023. As of February 20, 2024:
https://tomforb.es/blog/aws-keys-on-pypi/

Fortra, "Software for Adversary Simulations and Red Team Operations," webpage, undated. As of February 15, 2024:
https://www.cobaltstrike.com/

Franzen, Carl, "Mistral CEO Confirms 'Leak' of New Open Source AI Model Nearing GPT-4 Performance," *GamesBeat*, January 31, 2024.

Freedom House, *China: Transnational Repression Origin Country Case Study*, 2021.

Freund, Andres, "Backdoor in Upstream xz/liblzma Leading to SSH Server Compromise," memorandum to OSS Security, March 29, 2024.

Gallagher, Sean, "Iranian Hackers Used Visual Basic Malware to Wipe Vegas Casino's Network," *Ars Technica*, December 11, 2014.

Gallagher, Sean, "Playing NSA, Hardware Hackers Build USB Cable That Can Attack," *Ars Technica*, January 20, 2015.

Gartenberg, Chaim, "Security Startup Verkada Hack Exposes 150,000 Security Cameras in Tesla Factories, Jails, and More," *The Verge*, March 9, 2021.

Gatlan, Sergiu, "GitHub Rolls Out Free Secret Scanning for All Public Repositories," *Bleeping Computer*, December 15, 2022.

Gelb, Yehuda, Jossef Harush Kadouri, and Tzachi Zornshtain, "PyPi Is Under Attack: Project Creation and User Registration Suspended," Checkmarx, blog post, March 28, 2024. As of April 30, 2024:
https://checkmarx.com/blog/pypi-is-under-attack-project-creation-and-user-registration-suspended/

Gellman, Barton, and Ellen Nakashima, "U.S. Spy Agencies Mounted 231 Offensive Cyber-Operations in 2011, Documents Show," *Washington Post*, August 30, 2013.

Genkin, Daniel, Adi Shamir, and Eran Tromer, "RSA Key Extraction via Low-Bandwidth Acoustic Cryptanalysis," *CRYPTO 2014*, part I, LNCS 8616, Springer, 2014.

Genkin, Daniel, Lev Pachmanov, and Itamar Pipman, "ECDH Key-Extraction via Low-Bandwidth Electromagnetic Attacks on PCs," *RSA Conference Cryptographers' Track (CT-RSA) 2016*, LNCS 9610, Springer, 2016.

Gibbs, Samuel, "Dropbox Hack Leads to Leaking of 68m User Passwords on the Internet," *The Guardian*, August 31, 2016.

Glinsky, Albert, *Theremin: Ether Music and Espionage*, University of Illinois Press, 2005.

Glover, Claudia, "Pegasus Airline Breach Sees 6.5TB of Data Left in Unsecured AWS Bucket," *TechMonitor30*, August 17, 2022.

GoFetch, homepage, undated. As of April 30, 2024:
https://gofetch.fail/

Goodin, Dan, "Masked Thieves Storm into Chicago Colocation (Again!)," *The Register*, November 2, 2007.

Goodin, Dan, "Flame Malware Wielded Rare 'Collision' Crypto Attack Against Microsoft," *Ars Technica*, June 5, 2012.

Goodin, Dan, "Meet 'badBIOS,' the Mysterious Mac and PC Malware That Jumps Airgaps," *Ars Technica*, October 31, 2013.

Goodin, Dan, "NSA Could Put Undetectable 'Trapdoors' in Millions of Crypto Keys," *Ars Technica*, October 11, 2016.

Goodin, Dan, "Intel Patches Remote Hijacking Vulnerability That Lurked in Chips for 7 Years," *Ars Technica*, May 1, 2017.

Goodin, Dan, "Lapsus$ and SolarWinds Hackers Both Use the Same Old Trick to Bypass MFA," *Ars Technica*, March 29, 2022.

Goodin, Dan, "LastPass Says Employee's Home Computer Was Hacked and Corporate Vault Taken," *Ars Technica*, February 7, 2023a.

Goodin, Dan, "Latest Attack on PyPI Users Shows Crooks Are Only Getting Better," *Ars Technica*, February 14, 2023b.

Goodin, Dan, "Firmware Vulnerabilities in Millions of Computers Could Give Hackers Superuser Status," *Ars Technica*, July 20, 2023c.

Goodin, Dan, "Microsoft Signing Keys Keep Getting Hijacked, to the Delight of Chinese Threat Actors," *Ars Technica*, August 25, 2023d.

Google Project Zero, 0-Days In-the-Wild, database, undated. As of February 15, 2024:
https://googleprojectzero.github.io/0days-in-the-wild/

Gostev, Alexander, "Agent.btz: A Source of Inspiration?" Kaspersky SecureList, March 12, 2014.

Government of the United Kingdom, "The Bletchley Declaration by Countries Attending the AI Safety Summit, 1–2 November 2023," webpage, November 1, 2023. As of January 9, 2024:
https://www.gov.uk/government/publications/ai-safety-summit-2023-the-bletchley-declaration/the-bletchley-declaration-by-countries-attending-the-ai-safety-summit-1-2-november-2023

Grander, Danny, and Liran Tal, "A Post-Mortem of the Malicious Event-Stream Backdoor," *Snyk Blog*, December 6, 2018. As of February 15, 2024:
https://snyk.io/blog/a-post-mortem-of-the-malicious-event-stream-backdoor/

Grassi, Paul A., James L. Fenton, Elaine M. Newton, Ray A. Perlner, Andrew R. Regenscheid, William E. Burr, Justin P. Richer, Naomi B. Lefkovitz, Jamie M. Danker, Yee-Yin Choong, Kristen K. Greene, and Mary F. Theofanos, *Digital Identity Guidelines: Authentication and Lifecycle Management*, National Institute of Standards and Technology, NIST Special Publication 800-63B, October 16, 2023.

Graz University of Technology, "Meltdown and Spectre," webpage, undated. As of February 15, 2024:
https://meltdownattack.com/

Greenberg, Andy, "Mind the Gap: This Researcher Steals Data with Noise, Light, and Magnets," *Wired*, February 7, 2018a.

Greenberg, Andy, "The Untold Story of NotPetya, the Most Devastating Cyberattack in History," *Wired*, August 22, 2018b.

Greenberg, Andy, "How a Cloud Flaw Gave Chinese Spies a Key to Microsoft's Kingdom," *Wired*, July 12, 2023.

Greenberg, Andy, and Lisk Feng, "The Hotel Room Hacker," *Wired*, August 2017.

Grimes, William, "Marcus Klingberg, Highest-Ranking Soviet Spy Caught in Israel, Dies at 97," *New York Times*, December 3, 2015.

Grush, Loren, "A US-Born NASA Scientist Was Detained at the Border Until He Unlocked His Phone," *The Verge*, February 12, 2017.

Guri, Mordechai, "Air-Gap Research," webpage, undated. As of May 8, 2024:
https://www.covertchannels.com/

Guri, Mordechai, Gabi Kedma, Assaf Kachlon, and Yuval Elovici, "AirHopper: Bridging the Air-Gap Between Isolated Networks and Mobile Phones Using Radio Frequencies," *arXiv* preprint 1411.0237, November 2, 2014.

Hak5, "O.MG Cable," webpage, undated-a. As of October 3, 2023:
https://shop.hak5.org/products/omg-cable

Hak5, "USB Rubber Ducky," webpage, undated-b. As of February 20, 2024:
https://shop.hak5.org/products/usb-rubber-ducky

Hakçıl, Ata (ignis-sec), and Oxflotus, "PWDB—New Generation of Password Mass-Analysis (Pwdb-Public)," GitHub, undated. As of February 20, 2024:
https://github.com/ignis-sec/Pwdb-Public

Halderman, J. Alex, Seth D. Schoen, Nadia Heninger, William Clarkson, William Paul, Joseph A. Calandrino, Ariel J. Feldman, Jacob Appelbaum, and Edward W. Felten, *Lest We Remember: Cold Boot Attacks on Encryption Keys*, Princeton University Center for Information Technology Policy, 2008. As of February 20, 2024:
https://citp.princeton.edu/our-work/memory/

Harang, Rich, "Securing LLM Systems Against Prompt Injection," NVIDIA, blog post, August 3, 2023. As of February 20, 2024:
https://developer.nvidia.com/blog/securing-llm-systems-against-prompt-injection/

"Harold 'Kim' Philby and the Cambridge Three," Nova Online, undated. As of February 20, 2024:
https://www.pbs.org/wgbh/nova/venona/dece_philby.html

Harrison, Joshua, Ehsan Toreini, and Maryam Mehrnezhad, "A Practical Deep Learning-Based Acoustic Side Channel Attack on Keyboards," *arXiv* preprint 2308.01074, August 2, 2023.

"The Heartbleed Bug," webpage, June 3, 2020. As of April 3, 2024:
https://heartbleed.com

"Heathrow Probe After 'Security Files Found on USB Stick,'" BBC News, October 29, 2017.

Houser, Greg, "The Cuckoo's Egg & How it Relates to Cybersecurity," *Exida*, blog post, February 2, 2023. As of February 20, 2024:
https://www.exida.com/blog/the-cuckoos-egg-how-it-relates-to-cybersecurity

Institute of Electrical and Electronics Engineers, "2nd IEEE Conference on Secure and Trustworthy Machine Learning," conference homepage, University of Toronto, April 9–11, 2024. As of February 20, 2024:
https://satml.org/#

International Society of Automation, "The World's Only Consensus-Based Automation and Control Systems Cybersecurity Standards," ISA/IEC 62443 Series of Standards, undated. As of April 30, 2024:
https://www.isa.org/standards-and-publications/isa-standards/isa-iec-62443-series-of-standards

Internet Crime Complaint Center, *Internet Crime Report 2021*, Federal Bureau of Investigation, 2021.

ISACA and Looking Glass, *State of Cybersecurity 2022: Global Update on Workforce Efforts, Resources and Cyberoperations*, 2022.

Itkin, Eyal, and Itay Cohen, "The Story of Jian—How APT31 Stole and Used an Unknown Equation Group 0-Day," Check Point Research, February 22, 2021.

Jaros, Stephanie L., Katlin J. Rhyner, Shannen M. McGrath, and Erik R. Gregory, *The Resource Exfiltration Project: Findings from DoD Cases, 1985–2017*, Defense Personnel and Security Research Center, Office of People Analytics, 2019.

Kan, Michael, "LastPass Employee Could've Prevented Hack with a Software Update," *PC*, March 3, 2023.

Kaspersky, "Evil Twin Attacks and How to Prevent Them," undated-a. As of February 20, 2024:
https://usa.kaspersky.com/resource-center/preemptive-safety/evil-twin-attacks

Kaspersky, "What Is a Whaling Attack?" webpage, undated-b. As of May 8, 2024:
https://usa.kaspersky.com/resource-center/definitions/what-is-a-whaling-attack

Kaspersky, "What Is SIM Swapping?" webpage, undated-c. As of April 3, 2024:
https://www.kaspersky.com/resource-center/threats/sim-swapping

Kaspersky, Eugene, "A Matter of Triangulation," *Kaspersky Daily*, June 1, 2023.

Keeper Security, "Workplace Password Habits Leave Organizations Vulnerable to Cyber Attacks," webpage, 2021. As of February 20, 2024:
https://www.keepersecurity.com/resources/workplace-password-habits.html

Kilkelly, James, "When the Billion Dollar Hard Drive Grows Legs," Manufacturing.net, August 31, 2015.

Knight, Will, "A New Attack Impacts Major AI Chatbots—and No One Knows How to Stop It," *Wired*, August 1, 2023.

Koerner, Brendan I., "Inside the Cyberattack That Shocked the US Government," *Wired*, October 23, 2016.

Kornbluh, Peter, and Carlos Osorio, *The CIA's 'Minerva' Secret*, Briefing Book #696, National Security Archive, February 11, 2020.

Kostyuk, Nadiya, and Susan Landau, "Dueling over Dual_EC_DRGB: The Consequences of Corrupting a Cryptographic Standardization Process," *Harvard Law School National Security Journal*, Vol. 13, June 7, 2022.

Lakshmanan, Ravie, "Atlassian's Jira Service Management Found Vulnerable to Critical Vulnerability," *Hacker News*, February 3, 2023a.

Lakshmanan, Ravie, "NSO Group Used 3 Zero-Click iPhone Exploits Against Human Rights Defenders," *Hacker News*, April 20, 2023b.

Larin, Boris, "Operation Triangulation: The Last (Hardware) Mystery," Kaspersky SecureList, December 27, 2023.

Latent Space, "Commoditizing the Petaflop—with George Hotz of the Tiny Corp," webpage and video, June 20, 2023. As of May 8, 2024:
https://www.latent.space/p/geohot

Lazar, David, Haogang Chen, Xi Wang, and Nickolai Zeldovich, *Why Does Cryptographic Software Fail? A Case Study and Open Problems*, APSys '14, Association of Computer Machinery, June 25–26, 2014.

Lederer, Edith M., "UN Nuclear Chief: Ukraine Nuclear Plant Is 'Out of Control,'" Associated Press, August 3, 2022.

Ledger Academy, "Episode 3—Laser Fault Attacks," updated June 4, 2023. As of April 30, 2024:
https://www.ledger.com/academy/series/enter-the-donjon/episode-3-laser-fault-attacks

Lee, Dave, and Nick Kwek, "North Korean Hackers 'Could Kill,' Warns Key Defector," BBC News, May 29, 2015.

Lee, Micah, and Henrik Moltke, "Everybody Does It: The Messy Truth About Infiltrating Computer Supply Chains," *The Intercept*, January 24, 2019.

Lemos, Robert, "Inside Threat: Developers Leaked 10M Credentials, Passwords in 2022," *Dark Reading*, March 9, 2022.

Lenaert-Bergman, Bart, "What Are Downgrade Attacks?" *Crowdstrike*, March 14, 2023.

Lillis, Katie Bo, "CNN Exclusive: FBI Investigation Determined Chinese-Made Huawei Equipment Could Disrupt US Nuclear Arsenal Communications," CNN, July 25, 2022.

Limpalair, Christophe, "Hash Tables, Rainbow Table Attacks, and Salts," *Cybr*, July 11, 2022.

Lockheed Martin, "Cyber Kill Chain," webpage, undated. As of April 30, 2024:
https://www.lockheedmartin.com/en-us/capabilities/cyber/cyber-kill-chain.html

Lord, Bob, "Phishing Resistant MFA Is Key to Peace of Mind," Cybersecurity and Infrastructure Security Agency, blog post, April 12, 2023. As of February 20, 2024:
https://www.cisa.gov/news-events/news/phishing-resistant-mfa-key-peace-mind

Lumelsky, Avi, Guy Kaplan, and Gal Ebaz, "ShadowRay: First Known Attack Campaign Targeting AI Workloads Actively Exploited in the Wild," Oligo, March 26, 2024.

McCarthy, Rami (ramimac), "Background (aws-customer-security-incidents)," GitHub, undated. As of February 20, 2024:
https://github.com/ramimac/aws-customer-security-incidents/tree/main

McDaniel, Dwayne, "Toyota Suffered a Data Breach by Accidentally Exposing a Secret Key Publicly on GitHub," *GitGuardian*, October 11, 2022.

Metasploit, homepage, undated. As of February 15, 2024:
https://www.metasploit.com/

Microsoft, "Flame Malware Collision Attack Explained," blog post, June 6, 2012. As of February 20, 2024:
https://msrc.microsoft.com/blog/2012/06/flame-malware-collision-attack-explained/

Microsoft, "Failure Modes in Machine Learning," November 2, 2022.

"Microsoft AI Researchers Accidentally Expose 38 Terabytes of Confidential Data," *Hacker News*, September 19, 2023.

Microsoft Azure (azure), "counterfit," GitHub, undated. As of February 20, 2024:
https://github.com/Azure/counterfit/

Miller, Greg, and Ellen Nakashima, "WikiLeaks Says It Has Obtained Trove of CIA Hacking Tools," *Washington Post*, March 7, 2017.

Miller, Steve, Nathan Brubaker, Daniel Kapellmann Zafra, and Dan Caban, "Appendix B: Technical Analysis of Custom Attack Tools," in *TRITON Actor TTP Profile, Custom Attack Tools, Detections, and ATT&CK Mapping*, Mandiant, April 2019; via Internet Archive, stored on October 17, 2021. As of February 20, 2024:
https://web.archive.org/web/20211017032206/https:/www.fireeye.com/content/dam/fireeye-www/blog/pdfs/TRITON_Appendix_B.pdf

Miller, Steve, Nathan Brubaker, Daniel Kapellmann Zafra, and Dan Caban, *TRITON Actor TTP Profile, Custom Attack Tools, Detections, and ATT&CK Mapping*, Mandiant, updated November 25, 2022.

MITRE, "ATLAS," webpage, undated-a. As of February 15, 2024:
https://atlas.mitre.org/

MITRE, "ATT&CK," webpage, undated-b. As of February 15, 2024:
https://attack.mitre.org/

MITRE, "ATT&CK—Enterprise Matrix," webpage, undated-c. As of February 15, 2024:
https://attack.mitre.org/matrices/enterprise/

MITRE, "CVSS Scores Between 2013-07-30 and 2023-07-30," cvedetails.com, undated-c. As of February 15, 2024:
https://www.cvedetails.com/cvss-score-charts.php?fromform=1&vendor_id=&product_id=&startdate=2013-07-30&enddate=2023-07-30

MITRE, "Arbitrary Code Execution with Google Colab," incident date of July 2022a. As of February 15, 2024:
https://atlas.mitre.org/studies/AML.CS0018/

MITRE, "Compromised PyTorch Dependency Chain," incident date of December 25, 2022b. As of February 15, 2024:
https://atlas.mitre.org/studies/AML.CS0015/

MITRE, "Achieving Code Execution in MathGPT via Prompt Injection," incident date of January 28, 2023. As of February 15, 2024:
https://atlas.mitre.org/studies/AML.CS0016/

Moriuchi, Priscilla, and Bill Ladd, *China's Ministry of State Security Likely Influences National Network Vulnerability Publications*, Recorded Future, 2017.

Mukherjee, Subhabrata, Arindam Mitra, Ganesh Jawahar, Sahaj Agarwal, Hamid Palangi, and Ahmed Awadallah, "Orca: Progressive Learning from Complex Explanation Traces of GPT-4," *arXiv* preprint 2306.02707, June 5, 2023.

Naghibijouybari, H., A. Neupane, Z. Qian and N. Abu-Ghazaleh, "Side Channel Attacks on GPUs," *IEEE Transactions on Dependable and Secure Computing*, Vol. 18, No. 4, July–August 2021.

Nakashima, Ellen, and Craig Timberg, "NSA Officials Worried About the Day Its Potent Hacking Tool Would Get Loose. Then It Did," *Washington Post*, May 16, 2017.

National Counterintelligence and Security Center, *Technical Specifications for Construction and Management of Sensitive Compartmented Information Facilities*, version 1.5.1, July 26, 2021.

National Cyber Security Centre, "Reducing Data Exfiltration by Malicious Insiders," June 30, 2022.

National Insider Threat Task Force, *Insider Threat Program Maturity Framework*, 2018.

National Institute of Standards and Technology, "Cybersecurity Framework," webpage, undated. As of February 20, 2024:
https://www.nist.gov/cyberframework

National Institute of Standards and Technology, *Security Requirements for Cryptographic Modules*, Federal Information Processing Standards Publication 140-3, March 22, 2019.

National Institute of Standards and Technology, "Biden-Harris Administration Announces New NIST Public Working Group on AI," press release, June 22, 2023a.

National Institute of Standards and Technology, "NVD CWE Slice," National Vulnerability Database, August 3, 2023b. As of February 15, 2024:
https://nvd.nist.gov/vuln/categories

National Institute of Standards and Technology, *The NIST Cybersecurity Framework 2.0*, NIST CSWP 29 (Initial Public Draft), August 8, 2023c.

National Institute of Standards and Technology, "CVE-2023-7018 Detail," webpage, National Vulnerability Database, last modified December 29, 2023d. As of May 8, 2024:
https://nvd.nist.gov/vuln/detail/CVE-2023-7018

National Institute of Standards and Technology, "Secure Software Development Framework," January 23, 2024. As of February 15, 2024:
https://csrc.nist.gov/Projects/ssdf

National Intelligence Council, *Annual Threat Assessment of the U.S. Intelligence Community*, Office of the Director of National Intelligence, February 6, 2023.

Nevo, Sella, Dan Lahav, Ajay Karpur, Jeff Alstott, and Jason Matheny, "Securing Artificial Intelligence Model Weights: Interim Report," RAND Corporation, WR-A2849-1, 2023. As of May 8, 2024: https://www.rand.org/pubs/working_papers/WRA2849-1.html

NewAE Technology, "Glitching," webpage, undated. As of April 30, 2024: https://www.newae.com/glitching

Newman, Lily Hay, "AI Wrote Better Phishing Emails Than Humans in a Recent Test," *Wired*, August 7, 2021.

Newman, Lily Hay, "A Year Later, That Brutal Log4j Vulnerability Is Still Lurking," *Wired*, December 10, 2022.

Nicholls, Mark, "Types of Penetration Testing: Black Box, White Box & Grey Box," Redscan, December 10, 2023.

NIST—*See* National Institute of Standards and Technology.

NIST SP 800-171—*See* Ross, Ron, Victoria Pillitteri, Kelley Dempsey, Mark Riddle, and Gary Guissanie, *Protecting Controlled Unclassified Information in Nonfederal Systems and Organizations*, National Institute of Standards and Technology, NIST SP 800-171 Rev. 2, updated January 28, 2021.

"The NSA's Work to Make Crypto Worse and Better," *Ars Technica*, September 6, 2013. As of April 3, 2024: https://arstechnica.com/information-technology/2013/09/the-nsas-work-to-make-crypto-worse-and-better/

NVIDIA, "Product Security, webpage, undated. As of May 8, 2024: https://www.nvidia.com/en-us/security/

NVIDIA, *Confidential Compute on NVIDIA Hopper H100*, July 25, 2023.

OASIS, "STIX™ Version 2.0. Part 1: STIX Core Concepts— Committee Specification 01," webpage, July 19, 2017. As of May 8, 2024: https://docs.oasis-open.org/cti/stix/v2.0/stix-v2.0-part1-stix-core.html

O'Donnell, Lindsey, "Threatlist: IMAP-Based Attacks Compromising Accounts at 'Unprecedented Scale,'" *Threat Post*, March 14, 2019.

Oligo, "ShellTorch," webpage, undated. As of February 20, 2024: https://www.oligo.security/shelltorch

OpenAI, *Preparedness Framework (Beta)*, December 18, 2023.

OpenSSL, "Vulnerabilities," webpage, undated. As of February 15, 2024: https://www.openssl.org/news/vulnerabilities.html

Oprea, Alina, and Apostol Vassile, *Adversarial Machine Learning: A Taxonomy and Terminology of Attacks and Mitigations*, National Institute of Standards and Technology, NIST AI 100-2 E2023 (Initial Public Draft), March 8, 2023.

Oren, Yossi, "Oren Lab—Implementation Security and Side-Channel Attacks: Publications," webpage, undated. As of February 20, 2024: https://orenlab.sise.bgu.ac.il/Publications

Orland, Kyle, "MAME Devs Are Cracking Open Arcade Chips to Get Around DRM," *Ars Technica*, July 25, 2017.

Osgood, Rick, "Fast Hacks #6—Laser Spy Microphone," video, July 16, 2013. As of April 4, 2024: https://www.youtube.com/watch?v=K-96dX8ltO8

OWASP, "A02:2021—Cryptographic Failures," webpage, undated-a. As of February 20, 2024: https://owasp.org/Top10/A02_2021-Cryptographic_Failures/

OWASP, "OWASP Top Ten," webpage, undated-b. As of February 20, 2024: https://owasp.org/www-project-top-ten/

Page, Carly, "Rackspace Blames Ransomware Attack for Ongoing Exchange Outage," *TechCrunch*, December 6, 2022.

Park, Jangyong, Jaehoon Yoo, Jaehyun Yu, Jiho Lee, and Jae Seung Song, "A Survey on Air-Gap Attacks: Fundamentals, Transport Means, Attack Scenarios and Challenges," *Sensors*, Vol. 23, No. 6, 2023.

Patel, Dylan, and Gerald Wong, "GPT-4 Architecture, Infrastructure, Training Dataset, Costs, Vision, MoE," *SemiAnalysis*, July 10, 2023.

Perrin, Léo, "Partitions in the S-Box of Streebog and Kuznyechik," Paper 2019/092, *FSE 2019*, International Association for Cryptologic Research, 2019.

Petrizio, Andy, "Why the DVD Hack Was a Cinch," *Wired*, November 2, 1999.

Physics arXiv Blog, "Hack Forces Air-Gapped Computers to Transmit Their Own Secret Data," *Discover*, July 29, 2022.

"PixieFail UEFI Flaws Expose Millions of Computers to RCE, DoS, and Data Theft," *Hacker News*, January 18, 2024.

Pomerleau, Mark, "What's in the $9.6B Cyber Budget Request?" *C4ISRNET*, March 14, 2019.

Porter, Christopher, ed., *Threat Horizons: April 2023 Threat Horizons Report*, Google Cloud Office of the CISO, April 2023.

Potkin, Fanny, and Poppy McPherson, "Insight: How Myanmar's Military Moved in on the Telecoms Sector to Spy on Citizens," Reuters, May 18, 2021.

Prado, Angelo, Neal Harris, and Yoel Gluck, "Breach Attack," webpage, undated. As of April 3, 2024: https://www.breachattack.com

Privacy Rights Clearinghouse, "Data Breach Chronology," webpage, undated. As of February 20, 2024: https://privacyrights.org/data-breaches

ProofPoint, *2022 Social Engineering Report*, 2022.

Qualys, "SSL Pulse," webpage, February 2, 2024. As of February 15, 2024: https://ssllabs.com/ssl-pulse/

Ray, Siladitya, "Social Engineering: How a Teen Hacker Allegedly Managed to Breach Both Uber and Rockstar Games," *Forbes*, September 20, 2022.

RedLock CSI Team, "Lessons from the Cryptojacking Attack at Tesla," RedLock, February 20, 2018 As of March 28, 2024:
https://web.archive.org/web/20180221004714/https://blog.redlock.io/cryptojacking-tesla

Ross, Ron, Victoria Pillitteri, Kelley Dempsey, Mark Riddle, and Gary Guissanie, *Protecting Controlled Unclassified Information in Nonfederal Systems and Organizations*, National Institute of Standards and Technology, NIST SP 800-171 Rev. 2, updated January 28, 2021.

Sadowski, James, "Zero Tolerance: More Zero-Days Exploited in 2021 Than Ever Before," Mandiant blog post, April 21, 2022. As of February 15, 2024:
https://www.mandiant.com/resources/blog/zero-days-exploited-2021

Sanmillan, Ignacio, "Ramsay: A Cyber-Espionage Toolkit Tailored for Air-Gapped Networks," We Live Security, May 13, 2020.

Schneier, Bruce, "The Legacy of DES," *Schneier on Security*, blog post, October 6, 2004.

Schneier, Bruce, "Intentional Flaw in GPRS Encryption Algorithm GEA-1," *Schneier on Security*, blog post, June 17, 2021.

Schwartau, Winn, *Information Warfare*, 2nd ed., Thunder's Mouth Press, 1996.

Shane, Scott, Nicole Perlroth, and David E. Sanger, "Security Breach and Spilled Secrets Have Shaken the N.S.A. to Its Core," *New York Times*, November 12, 2017.

Sharma, Shweta, "Frequent Critical Flaws Open MLFlow Users to Imminent Threats," *CSO*, January 18, 2024.

"Shutting Out SamSam Ransomware," *Sophos News*, May 2, 2018.

Sophos, *SamSam: The (Almost) Six Million Dollar Ransomware*, 2018.

Sotirov, Alexander, Marc Stevens, Jacob Appelbaum, Arjen Lenstra, David Molnar, Dag Arne Osvik, and Benne de Weger, "MD5 Considered Harmful Today: Creating a Rogue CA Certificate," Eindhoven University of Technology, Mathematics and Computer Science, December 30, 2008. As of February 20, 2024:
https://www.win.tue.nl/hashclash/rogue-ca/

Southan, Jenny, "A Faked Master Key Gives Hackers Access to Millions of Hotel Rooms," *Wired*, April 25, 2018.

Spring, Tom, "Air-Gap Attack Turns Memory Modules into Wi-Fi Radios," *Threat Post*, December 17, 2020.

Stahie, Silviu, "Over 500,000 Credentials for Telnet Exposed IoT Devices and Servers Leaked Online," *Bitdefender*, January 20, 2020.

State of California Department of Justice, California Consumer Privacy Act (CCPA), May 10, 2023.

Stone, Maddie, "The Ups and Downs of 0-Days," Google Threat Analysis Group blog post, July 27, 2023. As of February 15, 2024:
https://blog.google/threat-analysis-group/0-days-exploited-wild-2022/

Sullivan, Nick, "Padding Oracles and the Decline of CBC-Mode Cipher Suites," Cloudflare, blog post, February 12, 2016.

Supply-chain Levels for Software Artifacts, homepage, undated. As of May 8, 2024:
https://slsa.dev

Synopsys, "[2023] Open Source Security and Risk Analysis Report," webpage, undated. As of February 15, 2024:
https://www.synopsys.com/software-integrity/resources/analyst-reports/open-source-security-risk-analysis.html

TEMPEST: A Signal Problem, National Security Agency, September 27, 2007.

Temple-Raston, Dina, "A 'Worst Nightmare' Cyberattack: The Untold Story of the SolarWinds Hack," *All Things Considered*, NPR, April 16, 2021.

TensorFlow, "Using TensorFlow Securely," GitHub, undated. As of February 15, 2024:
https://github.com/tensorflow/tensorflow/security/policy#vulnerabilities-in-tensorflow

Theohary, Catherine A., *Iranian Offensive Cyber Attack Capabilities*, Congressional Research Service, IF11406, January 13, 2020.

"Thousands of Android Apps Leak Hard-Coded Secrets, Research Shows," *Cybernews*, September 1, 2022.

"Top 7 Cybersecurity Threats: #3 Supply Chain Attacks," *MxD*, October 11, 2022.

Toulas, Bill, "CASPER Attack Steals Data Using Air-Gapped Computer's Internal Speaker," *Bleeping Computer*, March 12, 2023.

Tramèr, Florian, Fan Zhang, Ari Juels, Michael K. Reiter, and Thomas Ristenpart, *Stealing Machine Learning Models via Prediction APIs*, 25th USENIX Security Symposium, August 10–12, 2016.

Tromer, Eran, "LEISec: Laboratory for Experimental Information Security," webpage, undated. As of February 20, 2024:
https://cs-people.bu.edu/tromer/leisec.html

"Two LAPSUS$ Hackers Convicted in London Court for High-Profile Tech Firm Hacks," *Hacker News*, August 25, 2023.

Unsaflok, homepage, undated. As of April 30, 2024:
https://unsaflok.com/

United Nations General Assembly, 78th session, Seizing the Opportunities of Safe, Secure and Trustworthy Artificial Intelligence Systems for Sustainable Development, A/78/L.49, March 11, 2024.

U.S. Department of Justice, "Chinese Telecommunications Conglomerate Huawei and Subsidiaries Charged in Racketeering Conspiracy and Conspiracy to Steal Trade Secrets," press release, February 13, 2020.

U.S. Navy, "Emanations Security," in *Automated Information Systems Security Guidelines*, 1988; archived March 30, 2008. As of February 20, 2024:
https://web.archive.org/web/20080330004500/http:/www.cs.nps.navy.mil/curricula/tracks/security/AISGuide/navch16.txt

U.S. Senate, Committee on Commerce, Science, and Transportation, *A "Kill Chain": Analysis of the 2013 Target Data Breach*, majority staff report, 2014.

van Eck, Wim, "Electromagnetic Radiation from Video Display Units: An Eavesdropping Risk?" *Computers & Security*, Vol. 4, No. 4, December 1985.

Vijayan, Jaikumar, "One Year Later: Five Lessons Learned from the VA Data Breach," *Computer World*, June 1, 2007.

Vijayan, Jaikumar, "Target Attack Shows Danger of Remotely Accessible HVAC Systems," *Computer World*, February 7, 2014.

Vincent, James, "Meta's Powerful AI Language Model Has Leaked Online—What Happens Now?" *The Verge*, March 8, 2023.

Viswa, Chaitanya Adabala, Joachim Bleys, Eoin Leydon, Bhavik Shah, and Delphine Zurkiya, *Generative AI in the Pharmaceutical Industry: Moving from Hype to Reality*, McKinsey & Company, January 9, 2024.

Warren, Peter, and Michael Streeter, "Mission Impossible at the Sumitomo Bank," *The Register*, April 13, 2005.

Weulen Kranenbarg, Marleen, Thomas J. Holt, and Jeroen van der Ham, "Don't Shoot the Messenger! A Criminological and Computer Science Perspective on Coordinated Vulnerability Disclosure," *Crime Science*, Vol. 7, November 19, 2018.

"What Made the American Turncoat Tick?" CNN, May 10, 2002.

White House, *National Insider Threat Policy*, 2011.

White House, "Biden-Harris Administration Secures Voluntary Commitments from Leading Artificial Intelligence Companies to Manage the Risks Posed by AI," fact sheet, July 21, 2023.

Whittaker, Zack, "GoDaddy Says Data Breach Exposed over a Million User Accounts," *TechCrunch*, November 22, 2021.

Whittaker, Zack, "Okta Says Hundreds of Companies Impacted by Security Breach," *TechCrunch*, March 23, 2022.

Whittaker, Zack, "Parsing the UK Electoral Register Cyberattack," *TechCrunch*, August 9, 2023.

Williams, Brad D., "China's New Data Security Law Will Provide It Early Notice of Exploitable Zero Days," *Breaking Defense*, September 1, 2021.

Wiz, "ChaosDB," webpage, undated. As of February 20, 2024:
https://chaosdb.wiz.io/

Wolff, Josephine, "How a 2011 Hack You've Never Heard of Changed the Internet's Infrastructure," *Slate*, December 21, 2016.

Wray, Christopher, "Director's Opening Statement to the House Committee on Appropriations Subcommittee on Commerce, Justice, Science, and Related Agencies," testimony, Federal Bureau of Investigation, April 27, 2023.

Zerodium, homepage, undated-a. As of February 15, 2024:
https://zerodium.com/

Zerodium, "Zerodium Exploit Acquisition Program," webpage, undated-b. As of February 15, 2024:
https://zerodium.com/program.html

Zetter, Kim, "'Google' Hackers Had Ability to Alter Source Code," *Wired*, March 3, 2010.

Zetter, Kim, "Inside the Cunning, Unprecedented Hack of Ukraine's Power Grid," *Wired*, March 3, 2016.

Zetter, Kim, "Code Kept Secret for Years Reveals Its Flaw—a Backdoor," *Wired*, July 24, 2023.

Zetter, Kim, and Huib Modderkolk, "Revealed: How a Secret Dutch Mole Aided the U.S.-Israeli Stuxnet Cyberattack on Iran," *Yahoo News*, September 2, 2019.

"Zoom ZTP & AudioCodes Phones Flaws Uncovered, Exposing Users to Eavesdropping," *Hacker News*, August 12, 2023.